Memoirs
of the
American Mathematical Society

Volume 239 • Number 1129 (first of 6 numbers) • January 2016

On the Singular Set of Harmonic Maps into DM-Complexes

Georgios Daskalopoulos
Chikako Mese

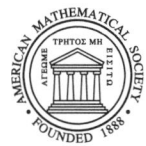

ISSN 0065-9266 (print)   ISSN 1947-6221 (online)

American Mathematical Society
Providence, Rhode Island

Library of Congress Cataloging-in-Publication Data

Daskalopoulos, Georgios, 1963-
  On the singular set of harmonic maps into DM-complexes / Georgios Daskalopoulos, Chikako Mese.
    pages cm. – (Memoirs of the American Mathematical Society, ISSN 0065-9266 ; volume 239, number 1129)
    Includes bibliographical references.
    ISBN 978-1-4704-1460-3 (alk. paper)
    1. Harmonic maps. 2. Differentiable manifolds. I. Mese, Chikako, 1968- II. Title.
QA614.73.D37   2016
514′.74–dc23   2015033756

DOI: http://dx.doi.org/10.1090/memo/1129

Memoirs of the American Mathematical Society

This journal is devoted entirely to research in pure and applied mathematics.

Subscription information. Beginning with the January 2010 issue, *Memoirs* is accessible from www.ams.org/journals. The 2016 subscription begins with volume 239 and consists of six mailings, each containing one or more numbers. Subscription prices for 2016 are as follows: for paper delivery, US$890 list, US$712.00 institutional member; for electronic delivery, US$784 list, US$627.20 institutional member. Upon request, subscribers to paper delivery of this journal are also entitled to receive electronic delivery. If ordering the paper version, add US$10 for delivery within the United States; US$69 for outside the United States. Subscription renewals are subject to late fees. See www.ams.org/help-faq for more journal subscription information. Each number may be ordered separately; *please specify number* when ordering an individual number.

Back number information. For back issues see www.ams.org/backvols.

Subscriptions and orders should be addressed to the American Mathematical Society, P. O. Box 845904, Boston, MA 02284-5904 USA. *All orders must be accompanied by payment.* Other correspondence should be addressed to 201 Charles Street, Providence, RI 02904-2294 USA.

Copying and reprinting. Individual readers of this publication, and nonprofit libraries acting for them, are permitted to make fair use of the material, such as to copy select pages for use in teaching or research. Permission is granted to quote brief passages from this publication in reviews, provided the customary acknowledgment of the source is given.

Republication, systematic copying, or multiple reproduction of any material in this publication is permitted only under license from the American Mathematical Society. Permissions to reuse portions of AMS publication content are handled by Copyright Clearance Center's RightsLink® service. For more information, please visit: http://www.ams.org/rightslink.

Send requests for translation rights and licensed reprints to reprint-permission@ams.org.

Excluded from these provisions is material for which the author holds copyright. In such cases, requests for permission to reuse or reprint material should be addressed directly to the author(s). Copyright ownership is indicated on the copyright page, or on the lower right-hand corner of the first page of each article within proceedings volumes.

Memoirs of the American Mathematical Society (ISSN 0065-9266 (print); 1947-6221 (online)) is published bimonthly (each volume consisting usually of more than one number) by the American Mathematical Society at 201 Charles Street, Providence, RI 02904-2294 USA. Periodicals postage paid at Providence, RI. Postmaster: Send address changes to Memoirs, American Mathematical Society, 201 Charles Street, Providence, RI 02904-2294 USA.

© 2015 by the American Mathematical Society. All rights reserved.
This publication is indexed in *Mathematical Reviews*®, *Zentralblatt MATH*, *Science Citation Index*®, *Science Citation Index*^{TM}*-Expanded*, *ISI Alerting Services*^{SM}, *SciSearch*®, *Research Alert*®, *CompuMath Citation Index*®, *Current Contents*®*/Physical, Chemical & Earth Sciences*. This publication is archived in *Portico* and *CLOCKSS*.
Printed in the United States of America.

∞ The paper used in this book is acid-free and falls within the guidelines
established to ensure permanence and durability.
Visit the AMS home page at http://www.ams.org/

10 9 8 7 6 5 4 3 2 1   20 19 18 17 16 15

Contents

Chapter 1.	Introduction	1
Chapter 2.	Harmonic maps into NPC spaces and DM-complexes	5
Chapter 3.	Regular and Singular points	9
Chapter 4.	Metric estimates near a singular point	15
Chapter 5.	Assumptions	19
Chapter 6.	The Target Variation	25
Chapter 7.	Lower Order Bound	37
Chapter 8.	The Domain variation	45
Chapter 9.	Order Function	55
Chapter 10.	The Gap Theorem	63
Chapter 11.	Proof of Theorems 1–4	67
Appendix A.	Appendix 1	75
Appendix B.	Appendix 2	83
Bibliography		89

Abstract

We prove that the singular set of a harmonic map from a smooth Riemannian domain to a Riemannian DM-complex is of Hausdorff codimension at least two. We also explore monotonicity formulas and an order gap theorem for approximately harmonic maps. These regularity results have applications to rigidity problems examined in subsequent articles.

Received by the editor September 28, 2013.
Article electronically published on June 4, 2015.
DOI: http://dx.doi.org/10.1090/memo/1129
2010 *Mathematics Subject Classification.* 53C43, 58E20.
Supported by the Nation Science Foundation under Grant DMS-1308708 and the Simons Foundation.
The second author was supported by the National Science Foundation under Grant DMS-1105599.

©2015 American Mathematical Society

CHAPTER 1

Introduction

Harmonic map theory from Riemannian domains to singular spaces originate with the work of Gromov-Schoen [**GS**] and was subsequently extended in [**KS1**], [**KS2**] and also [**Jo**]. The motivating question comes from rigidity theory. More precisely, one would like to know that a harmonic map, under appropriate curvature assumptions on the domain and the target spaces, is totally geodesic or even constant. This is the famous Bochner method which has been extensively used in the case when the target space is a smooth manifold. Recall that the Bochner formula is a differential equation involving higher derivatives of the map and relies on the smooth structure of the Riemannian manifolds involved. Therefore, in order to utilize it in the singular setting, the key is to show that *harmonic maps into singular spaces are regular enough on a big open set*.

In the seminal work of Gromov and Schoen [**GS**], it is shown that this is in fact the case when the target space is an F-connected simplicial complex. Roughly speaking, a k-dimensional F-connected complex is an NPC (non-positively curved) Euclidean k-complex where any two adjacent cells lie on a maximal flat, i.e. an image of the Euclidean space $\mathbf{R}^k$ embedded isometrically and totally geodesicly in the complex. Examples of F-connected complexes are Euclidean buildings. The main technical result of [**GS**] is to show that a harmonic map u from a smooth Riemannian domain Ω to a k-dimensional F-connected complex Y locally maps into a Euclidean space outside a set of codimension at least 2, or in other words, that the singular set $\mathcal{S}(u)$ of u is at least of Hausdorff codimension 2. To investigate the singular points, they show the existence of the order function (sometimes also called the frequency function) associated with a harmonic map. For example, for a harmonic function $u : \Omega \to \mathbf{R}$, the value of the order function $Ord^u(x)$ is the order with which u attains its value $u(x)$ at x. Alternatively, it is the degree of the dominant homogeneous harmonic polynomial which approximates $u - u(x)$ near x.

The question of superrigidity has played an important role in Geometric Group Theory, and it is beyond the scope of this introduction to summarize all the results of the vast literature. The goal of this paper is to lay the foundational analytic work needed in order to study superrigidity questions beyond the work of Gromov-Schoen, in other words, for a class of spaces larger than Euclidean buildings. For this purpose we introduce the notion of **D**ifferentiable **M**anifold complex (or simply DM-complex). A DM-complex is a cell complex Y with *branching-DM structure* in the sense that any two adjacent cells lie in a DM, the image of a **D**ifferentiable **M**anifold isometrically embedded in Y. Such complexes are assumed to be NPC but they can have arbitrary Riemannian metrics on their DM's. Special cases of such complexes are Euclidean and hyperbolic buildings. However, most of the work presented in this paper generalizes to an even larger class of spaces called **D**ifferentiable

1

Manifold spaces, which roughly speaking are metric spaces which have a differentiable manifold structure on a big open set. An example of a DM-space other than DM-complexes is the Weil-Petersson completion of Teichmüller space near a boundary stratum, which is related to important superrigidity questions of the Mapping Class Group. This space will be explored in a sequel paper.

We now summarize the main results of this paper. Our first main theorem can be stated as follows:

THEOREM 1. *If $u : \Omega \to Y$ is a harmonic map from an n-dimensional Riemannian domain to a k-dimensional NPC DM-complex, then the singular set $\mathcal{S}(u)$ of u has Hausdorff co-dimension at least 2 in Ω; i.e.*

$$\dim_{\mathcal{H}}(\mathcal{S}(u)) \leq n - 2.$$

We also prove

THEOREM 2. *Let $u : \Omega \to Y$ be as in Theorem 1. For any compact subdomain Ω_1 of Ω, there exists a sequence of smooth functions $\{\psi_i\}$ with $\psi_i \equiv 0$ in a neighborhood of $\mathcal{S}(u) \cap \overline{\Omega_1}$, $0 \leq \psi_i \leq 1$ and $\psi_i(x) \to 1$ for all $x \in \Omega_1 \backslash \mathcal{S}(u)$ such that*

$$\lim_{i \to \infty} \int_{\Omega} |\nabla \nabla u| |\nabla \psi_i| \, d\mu = 0.$$

A harmonic map $u : \Omega \to Y$ into a k-dimensional DM-complex can be written locally near a singular point $x_0 \in \mathcal{S}(u)$ as $u = (V, v)$ where V is the non-singular component map that maps into a Euclidean space $\mathbf{R}^j$ and v is the singular component map that maps into a lower dimensional complex Y_2^{k-j}. We partition $\mathcal{S}(u)$ as $\bigcup \mathcal{S}_j(u)$ where j indicates the dimension of the target space $\mathbf{R}^j$ of V (see Definitions 12 and 14). When the target space Y is an F-connected complex, u maps into the product of $\mathbf{R}^j$ and Y_2^{k-j}, and both components V and v are harmonic maps. Therefore, the analysis of the singular set of u can be inductively reduced to the study of the singular set of v which maps into a lower dimensional complex. This is in fact how it is argued in [**GS**]. In the case when the target space is a general DM-complex, u locally maps into the *twisted product* of $\mathbf{R}^j$ and Y_2^{k-j} which we denote by $(\mathbf{R}^j \times Y_2^{k-j}, d_G)$. The maps V and v are thus only *approximately* harmonic. More significantly, the map v is the non-dominant term of $u = (V, v)$. This presents the major technical difficulty of the paper. In analyzing the singular set of v, we prove a general monotonicity formula to deduce the existence of the order function and the order gap theorem for the approximate case. Here, we summarize our results:

THEOREM 3 (The Order of the Singular Component). *If $u : \Omega \to Y$ is a harmonic map from an n-dimensional Riemannian domain to a k-dimensional NPC DM-complex, $j \in \{0, \ldots, \min\{n, k\}\}$, $x_0 \in \mathcal{S}_j(u)$ and $u = (V, v)$ as above near x_0, then*

$$Ord^v(x_0) := \lim_{\sigma \to 0} \frac{\sigma E_{x_0}^v(\sigma)}{I_{x_0}^v(\sigma)}$$

exists. (See (1) for the notation.)

As with the case when v is harmonic, the main ingredient in proving the existence of the order function is a monotonicity formula. For this, the major steps are proving a *target variation formula* and a *domain variation formula*. This is achieved in sections 6 and 8 respectively. In fact, it follows from earlier work (cf. [**Me**] and

[**DM1**]) that all necessary monotonicity can be deduced as a formal consequence of the domain and target variation formulas combined with a Poincare type inequality proved in Section 7. The existence of the order function implies

THEOREM 4 (The Gap Theorem). *Under the same assumptions as Theorem 3, there exists $\epsilon_0 > 0$ such that $Ord^v(x) \geq 1 + \epsilon_0$ for all $x \in \mathcal{S}_j(u)$ near x_0.*

In the follow-up article [**DMV**], we show how to employ the results of this paper in order to prove superrigidity for representations of lattices into new classes of groups not covered by [**GS**], for example isometry groups of hyperbolic buildings. In subsequent articles, we will apply our results to study rigidity questions of Teichmüller space and the mapping class group. This is the reason why, as the reader may notice, our notation is a little more cumbersome than needed for proving the main results of the paper. For example, we state our main assumptions in Section 5 and deduce everything from there. These assumptions hold for the more general class of DM-spaces (like Teichmüller space, for example) from which we can deduce properties like monotonicity and order almost immediately.

Acknowledgement. The authors would like to thank Fang-Hua Lin and Bill Minicozzi for useful discussions.

CHAPTER 2

Harmonic maps into NPC spaces and DM-complexes

Let Ω be a smooth bounded n-dimensional Riemannian domain and (Y,d) a metric space. First recall that by the work of Gromov-Schoen and Korevaar-Schoen (cf. [**GS**] and [**KS1**]) one can define the Sobolev space of $W^{1,2}$ or *finite energy maps* $W^{1,2}(\Omega, Y) \subset L^2(\Omega, Y)$. In particular if $f \in W^{1,2}(\Omega, Y)$ one can define the energy density $|\nabla f|^2 \in L^1(\Omega)$ and the total energy

$$E^f = \int_\Omega |\nabla f|^2 d\mu$$

of f. Furthermore, it is shown in the references above that if $f \in W^{1,2}(\Omega, Y)$, then there exists a well-defined notion of a trace of f, denoted $Tr(f)$, which is an element of $L^2(\partial\Omega, Y)$. Two maps $f, g \in W^{1,2}(\Omega, Y)$ have the same trace (i.e. $Tr(f) = Tr(g)$) if and only if $d(f,g) \in W^{1,2}_0(\Omega)$. Given $x \in \Omega$ and f as above, we will use the following notation

(1) $$E^f_x(\sigma) := \int_{B_\sigma(x)} |\nabla f|^2 d\mu \quad \text{and} \quad I^f_x(\sigma) := \int_{\partial B_\sigma(x)} d^2(f, f(x)) d\Sigma.$$

DEFINITION 5. A $W^{1,2}$-map $u : \Omega \to Y$ to an NPC space Y is said to be *harmonic* or *energy minimizer* if, for any geodesic ball $B_r(x) \subset \Omega$, the restriction $f|_{B_r(x)}$ is energy minimizing among all $W^{1,2}$-maps with the same trace.

Let $u : \Omega \to Y$ be a harmonic map. By Section 1.2 of [**GS**], there exists a constant $c > 0$ depending only on the metric on Ω (in particular $c = 0$ when Ω is Euclidean) such that

$$\sigma \mapsto Ord^u(x, \sigma) := e^{c\sigma^2} \frac{\sigma \, E^u_x(\sigma)}{I^u_x(\sigma)}$$

is non-decreasing for any $x \in \Omega$. As a non-increasing limit of continuous functions,

$$Ord^u(x) := \lim_{\sigma \to 0} Ord^u(x, \sigma)$$

is an upper semicontinuous function. By following the proof of Theorem 2.3 in [**GS**], we see that $Ord^u(x) \geq 1$. The value $Ord^u(x)$ is called the order of u at x.

Fix $x_0 \in \Omega$ and choose a normal coordinate system centered at $x_0 = 0$. Set $\alpha := Ord^u(0)$. By Section 1.3 of [**GS**], there exists a constant $c > 0$ and $\sigma_0 > 0$ such that

$$\sigma \mapsto e^{c\sigma^2} \frac{I^u_0(\sigma)}{\sigma^{n-1+2\alpha}}$$

is monotone non-decreasing for $\sigma \in (0, \sigma_0)$. Thus,

(2) $$\lim_{\sigma \to 0} \mu_\sigma = 0$$

where

(3) $$\mu_\sigma := \sqrt{\frac{I_0^u(\sigma)}{\sigma^{n-1}}}.$$

Set $g_\sigma(x) = g(\sigma x)$ and define

$$u_\sigma : (B_1(0), g_\sigma) \to (Y, \mu_\sigma^{-1} d), \quad u_\sigma(x) = u(\sigma x).$$

By following Section 3 of [**GS**], we see that u_σ is a harmonic map with $E_0^{u_\sigma}(1) \leq 2\alpha$ and $I_0^{u_\sigma}(1) = 1$. Let $\delta = g(0)$ be the Euclidean metric defined by the value of g at 0. By Theorem 2.4.6 of [**KS1**], u_σ has a uniform modulus of continuity on compact sets independent of σ (with respect to the metric $g(0)$ on the domain which is uniformly equivalent to g_σ for σ small). By [**KS2**], Proposition 3.7 and a diagonalization argument, there exists $\sigma_i \to 0$ and a map $u_* : \mathbf{R}^n \to Y_*$ into an NPC space such that u_{σ_i} converges to u_* uniformly in the pull-back sense on every compact set. By (a slight modificaiton of) the L^2 trace theorem of [**KS1**], Theorem 1.12.2 and the fact that $I_0^{u_\sigma}(1) = 1$, we have that u_* is non-constant. Furthermore, by [**KS2**] Proposition 3.11 the energy of u_{σ_i} converges to u_* on compact subsets of $B_1(0)$. We claim that

(4) $\qquad\qquad u_*$ is an energy minimizer on $B_1(0)$.

Indeed, if $w : (B_1(0), g(0)) \to Y_*$ is an energy minimizing map with $w|_{\partial B_1(0)} = u_*|_{\partial B_1(0)}$, then Lemma 2.4.2 [**KS1**] implies that $d^2(u_*, w)$ is weakly subharmonic with zero boundary condition and hence $u_* = w$ on $B_1(0)$. Finally u_* is homogeneous degree α, i.e.

$$d(u_*(tx), u_*(0)) = t^\alpha d(u_*(x), u(0)) \text{ for } 0 \leq t \leq 1, \ x \in \mathbf{R}^n$$

by the same argument as in [**GS**] Proposition 3.3. Variations of the above argument will be used throughout the paper.

We now specialize to the case when Y is in a special class of cell complexes.

DEFINITION 6. Let $\mathbf{E}^d$ be an affine space. A convex piecewise linear polyhedron S with interior in some $\mathbf{E}^i \subset \mathbf{E}^d$ is called a cell. We will use the notation S^i to denote a cell S of dimension i. A *convex cell complex* or simply a *complex* Y in $\mathbf{E}^d$ is a finite collection $\mathcal{F} = \{S\}$ of cells satisfying the following properties: (i) the boundary ∂S^i of $S^i \in \mathcal{F}$ is a union of $T^j \in \mathcal{F}$ with $j < i$ (called the faces of S^i) and (ii) if $T^j, S^i \in \mathcal{F}$ with $j < i$ and $S^i \cap T^j \neq \emptyset$, then $T^j \subset S^i$.

For example, a simplicial complex is a cell complex whose cells are all simplices.

DEFINITION 7. A complex Y along with a metric $G = \{G^S\}$ is called a *Riemannian complex* if each cell S of Y is equipped with a smooth Riemannian metric G^S such that for each cell S, the component functions of G^S extend smoothly all the way to the boundary of S. Furthermore, if S' is a face of S then the restriction G^S to S' is equal to $G^{S'}$.

Throughout this paper, all cell complexes will have the additional property that *all cells are bounded* unless otherwise specified. If this is not the case, then we will write *unbounded cell complex*. Additionally, all cell complexes Y will be locally compact, Riemannian and NPC with respect to the distance function d induced from G^S.

DEFINITION 8. A k-dimensional Riemannian complex (Y, G) is said to have a *branching **D**ifferentiable **M**anifold structure* if given any two cells S_1 and S_2 of Y such that $S_1 \cap S_2 \neq \emptyset$, there exists a k-dimensional C^∞-differentiable, complete Riemannian manifold M and an isometric and totally geodesic embedding $J : M \to Y$ such that $S_1 \cup S_2 \subset J(M)$. Such complexes will be referred as DM-complexes. By an abuse of notation, we will often denote $J(M)$ by M and call it a DM (short for Differentiable Manifold).

REMARK 9. If any DM of a DM-complex is isometric to a k-dimensional Euclidean space, then the DM-complex is F-connected in the sense of [**GS**] Section 6.1. The NPC assumption implies that if M_1 and M_2 are DM's of a Riemannian DM-complex, then $M_1 \cap M_2$ is totally geodesic in M_1 and M_2.

Recall that for an arbitrary NPC space Y and a point $P \in Y$, the Alexandrov tangent cone $T_P Y$ of Y at P is the cone over the space of directions Π. Here, Π is the completion of the space of equivalence classes of geodesics emanating from P (where the equivalence relation $\sim$ is given by $\gamma_1 \sim \gamma_2 \Leftrightarrow$ the angle between γ_1, γ_2 at P is zero) along with the distance function defined by the angle at P. For a DM-complex Y, let C denote the tangent cone of Y at the point P as defined in [**Fe**] 3.1.21. Clearly, C is an unbounded cell complex and

(5) $$T_P Y \text{ is isometric to } (C, G(P))$$

where $G(P)$ is the metric defined by the value of G at P. Notice that if $P, Q \in int(S)$, then C for P and Q are isomorphic as sets. Let $\mathcal{M}_P$ be the set of all DM's passing through P. For each $M \in \mathcal{M}_P$, define $F_M = T_P M \subset C$. An immediate consequence is the following:

LEMMA 10. *If M is a DM in (Y, d_G), then F_M is a flat in $(C, G(P)) = T_P Y$. In particular, if Y is a DM-complex, then $T_P Y$ is F-connected in the sense of* [**GS**].

We can define the exponential map

(6) $$\exp_P^Y : T_P Y \to \bigcup_{M \in \mathcal{M}_P} M \subset Y$$

by piecing together the exponential maps defined on each $M \in \mathcal{M}_P$. This is equivalent to the exponential map defined from Alexandrov tangent cone point of view, i.e. given a unit speed geodesic γ and $t \in [0, \infty)$, $\exp_P^Y(\gamma, t) = \gamma(t)$.

Let $u : \Omega \to Y$ be a harmonic map into an NPC DM-complex and $x_0 \in \Omega$. By choosing normal coordinates, we can identify a neighborhood of $x_0 \in \Omega$ with a neighborhood of $0 \in \mathbf{R}^n$. Let $T_{u(x_0)} Y$ be the tangent cone of Y at $u(x_0)$. By a slight abuse of notation, we shall denote by

(7) $$G \text{ and } d_G \text{ respectively}$$

the pullback metric $\exp_{u(x_0)}^* G$ defined on C and the distance function induced by this pullback. Since we are only interested in the local behavior of u, we shall identify Y with (C, d_G). Let u_* be a tangent map of u at x_0. Recall that by definition, u_* is the limit (in the pullback sense as in [**KS2**] Section 3) of the maps

(8) $$u_{\sigma_i} : B_1(0) \to (C, \mu_{\sigma_i}^{-1} d_G), \quad u_{\sigma_i}(x) = u(\sigma_i x).$$

The induced pullback pseudodistances on $B_1(0)$ are the same as that of the maps

(9) $$\mu_{\sigma_i}^{-1} u_{\sigma_i} : B_1(0) \to (C, d_{G_{\sigma_i}}), \quad G_{\sigma_i}(y) = G(\mu_{\sigma_i} y).$$

The smoothness of the metric G implies that G_{σ_i} converges uniformly to the metric $G(u(0))$. Again, since $\mu_{\sigma_i}^{-1} u_{\sigma_i}$ have uniformly bounded energy $E_0^{\mu_{\sigma_i}^{-1} u_{\sigma_i}}(1)$ and uniformly bounded $I_0^{\mu_{\sigma_i}^{-1} u_{\sigma_i}}(1)$, we obtain by [**GS**] Theorem 2.4 and Arzela-Ascoli that $\mu_{\sigma_i}^{-1} u_{\sigma_i}$ converges locally uniformly to a limit map $u_0 : (B_1(0), g(0)) \to (C, d_{G(u(0))})$. By the equivalence of (8) and (9), u_0 must be equal to the tangent map u_*. We have thus shown

LEMMA 11. *Let $u : \Omega \to Y$ be a harmonic map into an NPC DM-complex. A tangent map of u at $x_0 \in \Omega$ is a homogeneous harmonic map into the NPC space $(C, d_{G(u(x_0))}) = T_{u(x_0)} Y$.*

CHAPTER 3

Regular and Singular points

As in the previous section, let Ω be an n-dimensional Riemannian domain and (Y, d_G) a k-dimensional NPC DM-complex.

DEFINITION 12. For a map $f : \Omega \to Y$, let $\hat{\mathcal{R}}(f)$ be the set of all points $x_0 \in \Omega$ such that for $\sigma_0 > 0$ sufficiently small

(10) $$f(B_{\sigma_0}(x_0)) \subset \exp^Y_{f(x_0)}(X_0)$$

where $X_0 \subset T_{u(x_0)}Y$ is isometric to $\mathbf{R}^k$. In particular, f maps a neighborhood of x_0 into a DM. If $u : \Omega \to Y$ is a harmonic map, a point $x_0 \in \Omega$ is called a *regular point* if $x_0 \in \hat{\mathcal{R}}(u)$ and $Ord^u(x_0) = 1$. A point $x_0 \in \Omega$ is called a *singular point* if it is not a regular point. Denote the set of regular points by $\mathcal{R}(u)$ and the set of singular points by $\mathcal{S}(u)$.

REMARK 13. The definition of a regular point in [**GS**] is slightly different than ours. Specifically, a regular point in [**GS**] may have order > 1 whereas ours does not.

DEFINITION 14. Let $u : \Omega \to Y$ be a harmonic map,
$$\mathcal{S}_0(u) - \{x_0 \in \Omega : Ord^u(x_0) > 1\},$$
$k_0 := min\{n, k\}$ and $\mathcal{S}_j(u) = \emptyset$ for $j \notin \{0, 1, \ldots, k_0\}$. For $j = 1, \ldots, k_0$, we define $\mathcal{S}_j(u)$ inductively as follows. Having defined $\mathcal{S}_m(u)$ for $m = j+1, \ldots, k_0+1$, define $\mathcal{S}_j(u)$ to be the set of points

$$x_0 \in \mathcal{S}(u) \backslash \left(\bigcup_{m=j+1}^{k_0} \mathcal{S}_m(u) \cup \mathcal{S}_0(u) \right)$$

with the property that there exists $\sigma_0 > 0$ such that

(11) $$u(B_{\sigma_0}(x_0)) \subset \exp^Y_{u(x_0)}(X_0)$$

where

(12) $$X_0 \subset T_{u(x_0)}Y \text{ is isometric to } \mathbf{R}^j \times Y_2^{k-j}$$

with Y_2^{k-j} a $(k-j)$-dimensional unbounded conical F-connected complex with vertex P_0. Set

$$\mathcal{S}_m^-(u) = \bigcup_{j=0}^m \mathcal{S}_j(u) \text{ and } \mathcal{S}_m^+(u) = \bigcup_{j=m}^k \mathcal{S}_j(u).$$

LEMMA 15. *The sets $\mathcal{S}_0(u), \mathcal{S}_1(u), \ldots, \mathcal{S}_{k_0-1}(u), \mathcal{S}_{k_0}(u)$ form a partition of $\mathcal{S}(u)$.*

PROOF. By definition, $\mathcal{S}_0(u), \ldots, \mathcal{S}_{k_0}(u)$ are mutually disjoint sets. Let $x_0 \in \mathcal{S}(u)$. If $Ord^u(x_0) > 1$, then $x_0 \in \mathcal{S}_0(u)$. If $Ord^u(x_0) = 1$, then the tangent map $u_* : \mathbf{R}^n \to T_{u(x_0)}Y$ at x_0 is a homogeneous degree 1 map and maps onto a flat $F_0 \subset T_{u(x_0)}Y$ by Proposition 3.1 of [**GS**]. Let X_0 be the union of all k-flats containing F_0. By Lemma 6.2 of [**GS**], X_0 is isometric to $\mathbf{R}^j \times Y_2^{k-j}$ where $j \in \{1, \ldots, k_0\}$ is the dimension of F_0. We can deduce from the proof of Lemma 6.2 of [**GS**] that Y_2^{k-j} is a cone. Furthermore, by the same lemma, u_* is effectively contained in X_0. Since $\sup_{B_r(x_0)} d(u, \exp^Y_{u(x_0)} \circ u_* \circ (\exp^{\Omega}_{x_0})^{-1}) \to 0$ as $r \to 0$, this implies by Theorem 5.1 of [**GS**] that $x_0 \in \mathcal{S}^+_j(u)$ and hence $x_0 \in \mathcal{S}_m(u)$ for some $m \in \{j, \ldots, k_0\}$. $\square$

LEMMA 16. *The sets $\mathcal{R}(u)$, $\mathcal{R}(u) \cup \mathcal{S}^+_m(u)$ are open and the sets $\mathcal{S}^-_m(u)$ are closed.*

PROOF. Clearly $\mathcal{R}(u)$ and $\mathcal{R}(u) \cup \mathcal{S}^+_0(u) = \Omega$ are open. Now assume $m > 0$ and $x_0 \in \mathcal{S}^+_m(u)$. Thus, $x_0 \in \mathcal{S}_j(u)$ for an integer $j \geq m$, hence $Ord^u(x_0) = 1$ and there exists $\sigma_0 > 0$ such that $u(B_{\sigma_0}(x_0)) \subset \exp^Y_{u(x_0)}(X_0)$ where X_0 is isometric to $\mathbf{R}^j \times Y_2^{k-j}$. Thus, $x \in B_\sigma(x_0)$ implies $x \in \mathcal{S}^l(u) \cup \mathcal{R}(u)$ for some $l \in \{j, \ldots, k_0\}$, i.e $x \in \mathcal{S}^+_m(u) \cup \mathcal{R}(u)$. This shows $\mathcal{S}^+_m(u) \cup \mathcal{R}(u)$ is open which in turn this implies $\mathcal{S}^-_m(u) = \Omega \backslash (\mathcal{S}^+_{m+1}(u) \cup \mathcal{R}(u))$ is closed. $\square$

Let $u : \Omega \to (Y, d_G)$ be a harmonic map and $x_\star \in \mathcal{S}_j(u)$ for $j > 0$. Thus, we can assume there exists $\sigma_\star > 0$ such that
$$u(B_{\sigma_\star}(x_\star)) \subset \exp^Y_{u(x_\star)}(\mathbf{R}^j \times Y_2^{k-j})$$
after isometrically identifying $\mathbf{R}^j \times Y_2^{k-j}$ with X_0 (cf. (10) and (12)). As seen by the proof of Lemma 15, $\mathbf{R}^j \times Y_2^{k-j}$ is the union of all k-flats $\{F_i\}_{i=1}^L$ containing the j-flat $\mathbf{R}^j \times \{P_0\}$, and we can write

(13) $$\mathbf{R}^j \times Y_2^{k-j} = \bigcup_{i=1}^L F_i.$$

Conversely, every k-flat of $\mathbf{R}^j \times Y_2^{k-j}$ is one of $\{F_i\}_{i=1}^L$. To see this, note that if F is a k-flat in $\mathbf{R}^j \times Y_2^{k-j}$ then $\pi_1(F)$ and $\pi_2(F)$ are flats in $\mathbf{R}^j$ and Y_2^{k-j} respectively where π_1 and π_2 are the projections onto the two factors $\mathbf{R}^j$ and Y_2^{k-j}. Since $\dim(\pi_1(F)) + \dim(\pi_2(F)) = \dim(F) = k$, we necessarily have $\dim(\pi_1(F)) = j$ and $\dim(\pi_2(F)) = k - j$. Thus, $\pi_1(F) = \mathbf{R}^j$, and since $\mathbf{R}^j \times Y_2^{k-j}$ is a cone, $\pi_2(F)$ must contain the point P_0. This implies that F contains the j-flat $\mathbf{R}^j \times \{P_0\}$.

We consider metrics

(14) $$G(u(x_\star)), \; G \text{ on } \mathbf{R}^j \times Y_2^{k-j} \text{ and } h \text{ on } Y_2^{k-j}$$

as follows. The flat metric $G(u(x_\star))$ is as in (5) with $P = u(x_\star)$. Notice that $G(u(x_\star))$ is a product metric on $\mathbf{R}^j \times Y_2^{k-j}$ by [**GS**] Lemma 6.2. The metric h is defined by restricting $G(u(x_\star))$ to Y_2^{k-j}. In particular, (Y_2^{k-j}, d_h) is a $(k-j)$-dimensional F-connected NPC complex. The metric G is the pullback metric via the exponentail map (6) as in (7). Note that then $(F_i, G|_{F_i})$ is a k-dimensional differentiable manifold for any F_i as in (13). Conversely, if $(M, G|_M)$ is a k-dimensional differentiable manifold containing $u(x_\star)$, then $(M, G(u(x_\star)))$ is isometric to $\mathbf{R}^k$, and hence $M = F_i$. In other words, $(\mathbf{R}^j \times Y_2^{k-j}, d_G)$ is a DM-complex where

$\{(F_i, G|_{F_i})\}$ is the set DM's of $(\mathbf{R}^j \times Y_2^{k-j}, d_G)$. We identify F_i with $\mathbf{R}^k$ such that $P_0 = (0, \ldots, 0) \in \mathbf{R}^{k-j}$. We will say that

(15) $$(\mathbf{R}^j \times Y_2^{k-j}, d_G) \text{ is a } \textit{local model}.$$

We are interested in the local properties of a harmonic map $u: \Omega \to Y$. Thus for $x_\star \in \Omega$ and $\sigma_\star > 0$ sufficiently small, we represent $u|_{B_{\sigma_\star}(x_\star)}$ as a harmonic map

(16) $$u = (V, v): (B_{\sigma_\star}(x_\star), g) \to (\mathbf{R}^j \times Y_2^{k-j}, d_G).$$

into a local model and refer to (16) as a *local representation*. Here, we assume that if we have the representation in the above form and $x_\star \in \mathcal{S}(u) \backslash \mathcal{S}_0(u)$, then $x \in \mathcal{S}_j(u)$ (cf. Definition 14). Furthermore, if $x_\star \in \mathcal{R}(u)$ then we assume $k = j$. The projection maps

$$V := \pi_1 \circ u : B_{\sigma_\star}(x_\star) \to \mathbf{R}^j \text{ and } v := \pi_2 \circ u : B_{\sigma_\star}(x_\star) \to Y_2^{k-j}$$

are called the the non-singular component and the singular component respectively. We will also need the following refined notion of regular.

DEFINITION 17. Let u as above, $x_0 \in B_{\sigma_\star}(x_\star)$, $\sigma_0 > 0$ such that $B_{\sigma_0}(x_0) \subset B_{\sigma_\star}(x_\star)$ and $w: (B_{\sigma_0}(x_0), g) \to (Y_2^{k-j}, d_h)$ be a harmonic map. A point $x \in \mathcal{R}(u)$ is said to be (u, w)-*regular* if there exists a flat F of Y_2^{k-j} and $r > 0$ such that $v(B_r(x)), w(B_r(x)) \subset F$. Denote by $\mathcal{R}(u, w)$ the set of all (u, w)-regular points.

LEMMA 18. *Let u and w as in Definition 17. For $x_0 \in \mathcal{R}(u) \cap \mathcal{R}(w)$, there exist $r > 0$ and a set Λ of finite $(n-1)$-Hausdorff measure such that $x \in \mathcal{R}(u, w)$ for any $x \in B_r(x_0) \backslash \Lambda$.*

PROOF. Let $\mathcal{F}$ denote the set of all $(k-j)$-flats of Y_2^{k-j}. Since $x_0 \in \mathcal{R}(u) \cap \mathcal{R}(w)$, there exist $r > 0$ and $F^v, F^w \in \mathcal{F}$ such that $v(B_r(x_0)) \subset F^v$ and $w(B_r(x_0)) \subset F^w$. For $F \in \mathcal{F} \backslash \{F^v\}$, there exists a finite set $\mathcal{L}_F^v$ of $(k-1)$-dimensional linear subspaces of F^v such that

$$\partial(F^v \cap F) \subset \bigcup_{L \in \mathcal{L}_F^v} L.$$

Intuitively speaking $\mathcal{L}_F^v$ is the set where flats can branch off F. Similarly define $\mathcal{L}_F^w$. We claim that for for every $L \in \mathcal{L}_F^v$, either (i) $v^{-1}(L) \cap B_r(x_0)$ is a real analytic subvariety of $B_r(x_0)$ of codimension at least 1 or (ii) $v(B_r(x_0)) \subset L$. We also claim an analogous statement for $L \in \mathcal{L}_F^w$ and $w^{-1}(L) \cap B_r(x_0)$. Since the proofs are similar, we only prove the first statement. First, isometrically identify F^v to $\mathbf{R}^{k-j}$ in such a way that if $(y^{j+1}, \ldots, y^k)$ are the standard coordinates of $\mathbf{R}^{k-j}$ then L is given by $\{(y^{j+1}, \ldots, y^k) : y^k = 0\}$. Let $(V, \ldots, u^k)$ be the coordinate expression of $u|_{B_r(x_0)} : B_r(x_0) \to \mathbf{R}^k \simeq \mathbf{R}^j \times F^v$. Since u satisfies the harmonic map equation, the unique continuation principle of elliptic p.d.e.'s implies that either $(u^k)^{-1}(0)$ is a subvariety of codimension at least 1 or $u^k \equiv 0$. This proves the claim. Let $\hat{\mathcal{L}}_F^v$ be the elements of $\mathcal{L}_F^v$ satisfying (i). Similarly define $\hat{\mathcal{L}}_F^w$. Then

$$\Lambda = \left(\bigcup_{F \in \mathcal{F} \backslash \{F^v\}} \bigcup_{L \in \hat{\mathcal{L}}_F^v} v^{-1}(L) \cup \bigcup_{F \in \mathcal{F} \backslash \{F^w\}} \bigcup_{L \in \hat{\mathcal{L}}_F^w} w^{-1}(L) \right) \cap B_r(x_0)$$

is clearly of finite $(n-1)$-Hausdorff measure. By construction, given any connected component C of $B_r(x_0) \backslash \Lambda$ and any $F \in \mathcal{F} \backslash \mathcal{F}^v$ either $v(C) \cap F = \emptyset$ or $v(C) \subset F$.

Hence (after assuming without loss of generality that the triangulation of Y^{k-j} has minimal number of cells), $v(C)$ is contained in a single closed k-cell, say S^v. Similarly, $w(C)$ is contained in a single (possibly the same) closed k-cell, say S^w. Since Y_2^{k-j} is F-connected and all cells are adjacent (containing P_0), there exists $F \in \mathcal{F}$ containing S^v and S^w. This shows $C \subset \mathcal{R}(u, w)$. □

COROLLARY 19. *If u and w as in Definition 17, then $B_r(x_0) \backslash \mathcal{R}(u, w)$ is of finite Hausdorff $(n-1)$-measure for any $r \in (0, \sigma_0)$.*

PROOF. Since $\mathcal{R}(w)$ is of Hausdorff codimension ≥ 2 by [**GS**], the assertion follows from Lemma 18. □

Let $x_0 \in \mathcal{S}_j(u)$ and identify $x_0 = 0$ via normal coordinates. Translating if necessary, assume $V(0) = 0$. Recall from (9) that the blow up maps of u at $x_0 = 0$ are the maps

$$u_\sigma(x) = (V_\sigma(x), v_\sigma(x)) := (\mu_\sigma^{-1} V(\sigma x), \mu_\sigma^{-1} v(\sigma x))$$

into $(\mathbf{R}^j \times Y_2^{k-j}, d_{G_\sigma})$ where $G_\sigma(y) = G(\mu_\sigma y)$. Also recall that the tangent map is a map into $(\mathbf{R}^j \times Y_2^{k-j}, d_{G(u(x_0))})$ by Lemma 11 and (12).

LEMMA 20. *If $u_* : (B_1(0), g(0)) \to (\mathbf{R}^j \times Y_2^{k-j}, d_{G(u(x_0))})$ is a tangent map of u at $x_0 \in \mathcal{S}_j(u)$, then $v_* := \pi_2 \circ u_* \equiv P_0$.*

PROOF. Assume on the contrary that $v_* \not\equiv P_0$. Since u_* is a homogeneous degree 1 map, so is v_*. By Proposition 3.1 of [**GS**] v_* maps into a flat F_0 of Y_2^{k-j} of dimension l. Let X_0 be the union of all k-flats containing F_0. By Lemma 6.2 of [**GS**], X_0 is isometric to $\mathbf{R}^{j+l} \times Z_2^{k-j-l}$ and u_* is effectively contained in $\mathbf{R}^{j+l} \times Z_2^{k-j-l}$. Since $\sup_{B_r(x)} d(u, \exp_{u(x)}^Y \circ u_* \circ (\exp_x^\Omega)^{-1}) \to 0$ as $r \to 0$, this implies that $x_0 \in \mathcal{S}_{j+l}^+(u)$ by Theorem 5.1 of [**GS**] which contradicts that $x_0 \in \mathcal{S}_j(u)$. □

Given a Lipschitz map

$$\hat{u} : (\hat{V}, \hat{v}) : (B_{\sigma_*}(x_*), g) \to (\mathbf{R}^j \times Y_2^{k-j}, d_G),$$

the component maps $\hat{V}$ and $\hat{v}$ can be seen as maps into a Riemannian manifold $(\mathbf{R}^j, H)$ where $H(V) = G(V, 0)$ and an NPC space (Y_2^{k-j}, d_h) respectively. We will prove later (cf. Lemma 29) that for a.e. $x \in B_{\sigma_*}(x_*)$

$$\left| |\nabla \hat{u}|^2(x) - \left(|\nabla \hat{V}|^2(x) + |\nabla \hat{v}|^2(x) \right) \right| \leq C d^2(\hat{v}(x), P_0) \tag{17}$$

where the constant C depends only on the Lipschitz constant of $\hat{u}$ and the constant in the estimates (28)-(32) for the target metric G. By an abuse of notation, we have used $|\cdot|$ to denote the norms with respect to d_H, d_h and d_G for maps into $\mathbf{R}^j$, Y_2^{k-j} and $\mathbf{R} \times Y_2^{k-j}$ respectively. For now, we assume this property and we obtain the following as a corollary of Lemma 20.

LEMMA 21. *Assume that the DM-complex $(\mathbf{R}^j \times Y_2^{k-j}, d_G)$ satisfies (17). If $u : (V, v) : (B_{\sigma_*}(x_*), g) \to (\mathbf{R}^j \times Y_2^{k-j}, d_G)$ is a harmonic map, then for a.e $x \in \mathcal{S}_j(u)$*

$$|\nabla v|^2(x) = 0 \quad and \quad |\nabla V|^2(x) = |\nabla u|^2(x).$$

3. REGULAR AND SINGULAR POINTS

PROOF. Since $|\nabla v|^2$ is L^1, almost every point of $B_{\sigma_*}(x_*)$ is a Lebesgue point. Let $x \in \mathcal{S}_j(u)$ be a Lebesgue point of $|\nabla v|^2$ and C be the Lipschitz bound of u in $\overline{B_r(x)} \subset B_{\sigma_*}(x_*)$. After identifying $x = 0$ via normal coordinates, let $u_{\sigma_i} = (V_{\sigma_i}, v_{\sigma_i})$ be a sequence blow up maps converging to a tangent map $u_* = (V_*, v_*)$. Then (17) implies

$$(18) \qquad E^{u_\sigma}(r) = \left(E^{V_\sigma}(r) + E^{v_\sigma}(r)\right) + O(\sigma^2).$$

Combined with Lemma 20, we obtain

$$(19) \qquad E^{u_*}(r) = E^{V_*}(r) + E^{v_*}(r) = E^{V_*}(r).$$

Therefore,

$$\begin{aligned}
\limsup_{i \to \infty} E^{V_{\sigma_i}}(r) &\leq \limsup_{i \to \infty} E^{V_{\sigma_i}}(r) + \limsup_{i \to \infty} E^{v_{\sigma_i}}(r) \\
&= \lim_{i \to \infty} E^{u_{\sigma_i}}(r) \quad \text{(by (18))} \\
&= E^{u_*}(r) \quad \text{(by [\textbf{KS2}] Theorem 3.11)} \\
&= E^{V_*}(r) \quad \text{(by (19))} \\
&\leq \liminf_{i \to \infty} E^{V_{\sigma_i}}(r)
\end{aligned}$$

where the last inequality is by the lower semicontinuity of energy [**KS2**] Lemma 3.8. This immediately implies

$$(20) \qquad \lim_{i \to \infty} E^{V_{\sigma_i}}(r) = \lim_{i \to \infty} E^{u_{\sigma_i}}(r) \quad \text{and} \quad \lim_{i \to \infty} E^{v_{\sigma_i}}(r) = 0.$$

Therefore,

$$\begin{aligned}
|\nabla v|^2(0) &= \lim_{i \to \infty} \frac{1}{Vol(B_{\sigma_i r}(0))} \int_{B_{\sigma_i r}(0)} |\nabla v|^2 d\mu \\
&= \lim_{i \to \infty} \frac{\mu_{\sigma_i}^2}{Vol(B_r(0))} \int_{B_r(0)} |\nabla v_{\sigma_i}|^2 d\mu_{\sigma_i} \\
&\leq \lim_{i \to \infty} \frac{C^2}{Vol(B_r(0))} \int_{B_r(0)} |\nabla v_{\sigma_i}|^2 d\mu_{\sigma_i} \\
&= 0 \quad \text{(by (20))}.
\end{aligned}$$

This implies the first assertion. The second follows immediately from the first and (19). □

CHAPTER 4

Metric estimates near a singular point

Given a harmonic map $u : \Omega \to (Y, d_G)$, the goal of this section is to derive some estimates of the metric near $u(x_\star)$ for $x_\star \in \mathcal{S}_j(u)$, $j > 0$. Thus, let $(\mathbf{R}^j \times Y_2^{k-j}, d_G)$ and (Y_2^{k-j}, d_h) be as in (14). We will denote by $V = (V^1, \ldots, V^j)$ the standard coordinates of $\mathbf{R}^j$, $v = (v^{j+1}, \ldots, v^k)$ the standard coordinates of $\mathbf{R}^{k-j}$ and (V, v) the standard coordinates of $\mathbf{R}^k = \mathbf{R}^j \times \mathbf{R}^{k-j}$.

We will first construct a coordinate chart for a DM M of $(\mathbf{R}^j \times Y_2^{k-j}, d_G)$ in a neighborhood of $(0, P_0)$. First, we identify $\mathbf{R}^j \times \{0\}$ with the lowest dimensional singular locus $\mathbf{R}^j \times \{P_0\} \subset M$ of $\mathbf{R}^j \times Y_2^{k-j}$ by the identity map. Next, let $\{e_{j+1}(V, 0), \ldots, e_k(V, 0)\}$ be an orthonormal frame of the normal space to $\mathbf{R}^j \times \{0\}$ in M. Furthermore, for each $V \in \mathbf{R}^j$, let $\Phi_V : \mathbf{R}^k \to M$ be a normal coordinate chart centered at $(V, 0)$ with

$$d\Phi_V\Big|_{T_{(V,0)}\mathbf{R}^k}\left(\frac{\partial}{\partial v^m}\right) = e_m(V, 0), \ \forall m = j+1, \ldots, k.$$

Finally, we construct coordinates for a neighborhood of $(0, 0) \in M$ by defining a diffeomorphism Φ that agrees with the normal coordinate chart Φ_V on the slice $\{V\} \times \mathbf{R}^{k-j}$. More precisely, for a sufficiently small neighborhood $\mathcal{U}$ of $(0, 0) \in \mathbf{R}^j \times \mathbf{R}^{k-j}$, define coordinates (V, v) via the coordinate chart

$$\Phi : \mathcal{U} \subset \mathbf{R}^j \times \mathbf{R}^{k-j} \to \Phi(\mathcal{U}) \subset M, \quad \Phi(V, v) = \Phi_V\big|_{\{0\} \times \mathbf{R}^{k-j}}(v).$$

We are only interested in the local properties of $(\mathbf{R}^j \times Y_2^{k-j}, d_G)$. Hence, by an abuse of notation, we will identify each DM M with $\mathbf{R}^j \times \mathbf{R}^{k-j}$ along with (the extension of) the pullback of the metric G via the coordinates (V, v) (which we shall still denote by G). In particular, since $\mathbf{R}^j \times Y_2^{k-j}$ is a union of k-flats $\{F_i\}$ and $(F_i, G|_{F_i})$ is a DM for each i (cf. (13)), we can express every point $P \in \mathbf{R}^j \times Y_2^{k-j}$ as $P = (V, v)$.

LEMMA 22. *Let* $M = (\mathbf{R}^j \times \mathbf{R}^{k-j}, G)$ *be a DM in* $(\mathbf{R}^j \times Y_2^{k-j}, d_G)$ *and let*

$$G = \begin{pmatrix} \mathbf{G}_{11}(V, v) & \mathbf{G}_{12}(V, v) \\ \mathbf{G}_{21}(V, v) & \mathbf{G}_{22}(V, v) \end{pmatrix}$$

be the matrix representation of G *with*

$$\mathbf{G}_{11}(V, v) = (G_{IJ}(V, v)) \quad \mathbf{G}_{12}(V, v) = (G_{Il}(V, v))$$
$$\mathbf{G}_{21}(V, v) = (G_{lI}(V, v)) \quad \mathbf{G}_{22}(V, v) = (G_{lm}(V, v))$$

for $I, J = 1, \ldots, j$ *and* $l, m = j+1, \ldots, k$. *Then for* (V, v) *sufficiently close to* $(0, 0)$, *there exists a constant* $C > 0$ *depending only on*

(21)   the sup norm of the second derivatives of the metric G,

such that

(22)
$$|G_{IJ}(V,v) - G_{IJ}(V,0)| \leq C|v|^2, \quad |\tfrac{\partial}{\partial v^i}G_{IJ}(V,v)| \leq C|v|$$
$$|G_{Il}(V,v)| \leq C|v|^2, \quad |\dot{G}_{Il}(V,v)| \leq C|v|$$
$$|G_{lm}(V,v) - \delta_{lm}| \leq C|v|^2, \quad |\dot{G}_{lm}(V,v)| \leq C|v|$$

In the above, $\dot{G}$ is used indicate any derivatives (i.e. $\frac{\partial}{\partial V^I}$ or $\frac{\partial}{\partial v^l}$) of G.

PROOF. To prove (22), we first verify the following equalities:

(i) $\quad \frac{\partial}{\partial V^J} < \frac{\partial}{\partial V^I}, \frac{\partial}{\partial v^l} > (V,0) = 0$

(ii) $\quad \frac{\partial}{\partial v^m} < \frac{\partial}{\partial V^I}, \frac{\partial}{\partial v^l} > (V,0) = 0$

(iii) $\quad \frac{\partial}{\partial v^m} < \frac{\partial}{\partial V^I}, \frac{\partial}{\partial V^J} > (V,0) = 0$

(iv) $\quad \frac{\partial}{\partial V^I} < \frac{\partial}{\partial v^l}, \frac{\partial}{\partial v^m} > (V,0) = 0$

(v) $\quad \frac{\partial}{\partial v^m} < \frac{\partial}{\partial v^l}, \frac{\partial}{\partial v^p} > (V,0) = 0.$

Indeed, since $\{e_m(V,0)\}_{m=j+1,\ldots,k}$ is an orthonormal frame of the normal space of $\mathbf{R}^j \times \{P_0\}$, we have that

$$< \frac{\partial}{\partial V^I}, \frac{\partial}{\partial v^l} > (V,0) \equiv 0 \text{ and } < \frac{\partial}{\partial v^l}, \frac{\partial}{\partial v^m} > (V,0) \equiv \delta_{lm}$$

which immediately implies (i) and (iv). We next verify (ii). Fix $(V_0, 0)$ and identify $(V_0, 0) = (0,0)$ for simplicity. Denoting the normal coordinates centered at $(0,0)$ by $(\tilde{V}, \tilde{v})$, we have

(23) $$\nabla_X \frac{\partial}{\partial \tilde{v}^m}(0,0) = 0, \ \forall X \in T_{(0,0)}\mathbf{R}^k, m = j+1,\ldots,k.$$

Since $v = \tilde{v}$ on the slice $\{0\} \times \mathbf{R}^{k-j}$ by the definition of Φ, we have

(24) $$\frac{\partial}{\partial v^m}(0,v) = \frac{\partial}{\partial \tilde{v}^m}(0,v).$$

Furthermore, $(V, \tilde{v}) \mapsto (V, v)$ is a diffeomorphism in a neighborhood of $(0,0)$, and hence $(V, \tilde{v})$ are also coordinates in a neighborhood of $(0,0)$. In particular, this implies that

(25) $$\nabla_{\frac{\partial}{\partial \tilde{v}^m}} \frac{\partial}{\partial V^I} = \nabla_{\frac{\partial}{\partial V^I}} \frac{\partial}{\partial \tilde{v}^m}.$$

Thus, we have at $(0,0)$

$$\frac{\partial}{\partial v^m} < \frac{\partial}{\partial V^I}, \frac{\partial}{\partial v^l} > = \frac{\partial}{\partial \tilde{v}^m} < \frac{\partial}{\partial V^I}, \frac{\partial}{\partial \tilde{v}^l} > \quad \text{by (24)}$$
$$= < \nabla_{\frac{\partial}{\partial \tilde{v}^m}} \frac{\partial}{\partial V^I}, \frac{\partial}{\partial \tilde{v}^l} > + < \frac{\partial}{\partial V^I}, \nabla_{\frac{\partial}{\partial \tilde{v}^m}} \frac{\partial}{\partial \tilde{v}^l} >$$
$$= < \nabla_{\frac{\partial}{\partial \tilde{v}^m}} \frac{\partial}{\partial V^I}, \frac{\partial}{\partial \tilde{v}^l} > \quad \text{by (23)}$$
$$= < \nabla_{\frac{\partial}{\partial V^I}} \frac{\partial}{\partial \tilde{v}^m}, \frac{\partial}{\partial \tilde{v}^l} > \quad \text{by (25)}$$
$$= 0 \quad \text{by (23)}$$

which proves (ii). Similarly for (iii) and (v), we have at $(0,0)$

$$\begin{aligned}
\frac{\partial}{\partial v^m}<\frac{\partial}{\partial V^I},\frac{\partial}{\partial V^J}> &= \frac{\partial}{\partial \tilde{v}^m}<\frac{\partial}{\partial V^I},\frac{\partial}{\partial V^J}> \\
&= <\nabla_{\frac{\partial}{\partial \tilde{v}^m}}\frac{\partial}{\partial V^I},\frac{\partial}{\partial V^J}>+<\frac{\partial}{\partial V^I},\nabla_{\frac{\partial}{\partial \tilde{v}^m}}\frac{\partial}{\partial V^J}> \\
&= <\nabla_{\frac{\partial}{\partial V^I}}\frac{\partial}{\partial \tilde{v}^m},\frac{\partial}{\partial V^J}>+<\frac{\partial}{\partial V^I},\nabla_{\frac{\partial}{\partial V^J}}\frac{\partial}{\partial \tilde{v}^m}> \\
&= 0
\end{aligned}$$

and

$$\begin{aligned}
\frac{\partial}{\partial v^m}<\frac{\partial}{\partial v^l},\frac{\partial}{\partial v^m}> &= \frac{\partial}{\partial \tilde{v}^m}<\frac{\partial}{\partial \tilde{v}^l},\frac{\partial}{\partial \tilde{v}^m}> \\
&= <\nabla_{\frac{\partial}{\partial \tilde{v}^m}}\frac{\partial}{\partial \tilde{v}^l},\frac{\partial}{\partial \tilde{v}^m}>+<\frac{\partial}{\partial \tilde{v}^l},\nabla_{\frac{\partial}{\partial \tilde{v}^m}}\frac{\partial}{\partial \tilde{v}^m}> \\
&= 0.
\end{aligned}$$

The estimates of (22) follow from the inequalities (i) through (v). Here, we will only prove

(26) $$|\mathbf{G}_{11}(V,v) - \mathbf{G}_{11}(V,0)| \leq C|v|^2$$

and

(27) $$|\frac{\partial}{\partial v^l}\mathbf{G}_{11}(V,v)| \leq C|v|$$

since the other estimates follow by a similar argument. To prove (26), first apply the Mean Value Theorem and the chain rule to obtain for some $\tau \in (0,1)$

$$\begin{aligned}
\mathbf{G}_{11}(V,v) - \mathbf{G}_{11}(V,0) &= \left(\frac{\partial}{\partial t}\mathbf{G}_{11}(V,tv)\right)\Big|_{t=\tau} \\
&= \sum_{m=j+1}^{k} v^m \frac{\partial}{\partial v^m}\mathbf{G}_{11}(V,\tau v).
\end{aligned}$$

Since (iii) implies

$$\frac{\partial}{\partial v^m}\mathbf{G}_{11}(V,0) = 0, \ \forall m = j+1,\ldots,k,$$

we have for some $\sigma \in (0,1)$

$$\begin{aligned}
\frac{\partial}{\partial v^m}\mathbf{G}_{11}(V,\tau v) &= \left(\frac{\partial}{\partial s}\left(\frac{\partial}{\partial v^m}\mathbf{G}_{11}(V,s\tau v)\right)\right)\Big|_{s=\sigma} \\
&= \sum_{l=j+1}^{k} \tau v^l \frac{\partial^2}{\partial v^l \partial v^m}\mathbf{G}_{11}(V,\sigma \tau v).
\end{aligned}$$

Together, we have

$$\mathbf{G}_{11}(V,v) - \mathbf{G}_{11}(V,0) = \sum_{l,m=j+1}^{k} \tau v^l v^m \frac{\partial^2}{\partial v^l v^m}\mathbf{G}_{11}(V,\sigma \tau v)$$

which implies (26) with C as in (21). To prove (27), we first note that $\frac{\partial}{\partial v^l}\mathbf{G}_{11}(V,0) = 0$ by (ii). Thus, for some $\tau \in (0,1)$

$$\begin{aligned}\frac{\partial}{\partial v^l}\mathbf{G}_{11}(V,v) &= \left(\frac{\partial}{\partial t}\left(\frac{\partial}{\partial v^l}\mathbf{G}_{11}(V,tv)\right)\right)\Big|_{t=\tau} \\ &= \sum_{l=j+1}^{k} v^l \frac{\partial^2}{\partial v^m \partial v^l}\mathbf{G}_{11}(V,\tau v)\end{aligned}$$

which implies (27) with C as in (21). $\square$

CHAPTER 5

Assumptions

In the subsequent chapters, we will analyze a local representation (cf. (16))
$$u = (V, v) : (B_{\sigma_\star}(x_\star), g) \to (\mathbf{R}^j \times Y_2^{k-j}, d_G)$$
of a harmonic map into a DM-complex where the component maps V and v can be seem as maps into a Riemannian manifold $(\mathbf{R}^j, H)$ where $H(V) = G(V, 0)$ and an NPC space (Y_2^{k-j}, d_h) respectively. In this section, we summarize all the notation and list the relevant properties that will be used. On the other hand, the DM-complexes are special cases of **D**ifferentiable **M**anifold spaces or simply DM-spaces that we will study in our forthcoming papers. For example, the Weil-Petersson completion of Teichmüller space is an example of a DM-space. In other words, we are interested in applying the results of this paper to a more general setting. For this reason, we state the properties of the metric space $(\mathbf{R}^j \times Y_2^{k-j}, d_G)$ and the harmonic map u in a general form (as assumptions) below.

ASSUMPTION 1. *The metric space (Y_2^{k-j}, d_h) is an NPC space with a homogeneous structure with respect to a base point $P_0 \in Y_2^{k-j}$. In other words, there is a continuous function*
$$\mathbf{R}_{>0} \times Y_2 \to Y_2, \ (\lambda, P) \mapsto \lambda P$$
such that $\lambda P_0 = P_0$ for every $\lambda > 0$ and the distance function d is homogeneous of degree 1, i.e.
$$d(\lambda P, \lambda P') = \lambda d(P, P'), \ \forall P, P' \in Y_2.$$

REMARK 23. In this paper, we are interested in the case where Y_2^{k-j} is a $(k-j)$-dimensional unbounded conical F-connected complex with vertex P_0. The homogeneous structure is given by the scalar multiplication in Euclidean space (after identifying the $(k-j)$-dimensional flat that contains P and P_0 with $\mathbf{R}^{k-j}$ such that P_0 is identified with the origin).

Recall the estimates of the metrics G and h in Lemma 22. We will state these estimates in in a general setup below.

ASSUMPTION 2. *The metric space $(\mathbf{R}^j \times Y_2^{k-j}, d_G)$ is an NPC space. The Riemannian metric H of $\mathbf{R}^j$ and the metric h of Y_2^{k-j} is such that on every DM $(\mathbf{R}^j \times F^{k-j}, G)$, the metric G is asymptotically the product metric*
$$G_0(V, v) = H(V) \oplus h(v).$$

By this we mean the following. There exist constants $C > 0$ and $\epsilon \in (0, \frac{1}{2})$ such that if, with respect to the standard coordinates $(V^1, \ldots, V^j)$ of $\mathbf{R}^j$ and some coordinates $(v^{j+1},\ldots,v^k)$ of F^{k-j}, at P_0 we have

$$H(V) = (H_{IL}(V)), \qquad H^{-1}(V) = (H^{IL}(V)),$$
$$h(v) = (h_{il}(v)), \qquad h^{-1}(v) = (h^{il}(v)),$$
$$G(V,v) = \begin{pmatrix} G_{IL}(V,v) & G_{Il}(V,v) \\ G_{iL}(V,v) & G_{il}(V,v) \end{pmatrix}, \quad G^{-1}(V,v) = \begin{pmatrix} G^{IL}(V,v) & G^{Il}(V,v) \\ G^{iL}(V,v) & G^{il}(V,v) \end{pmatrix}$$

with $I, L = 1, \ldots, j$ and $i, l = j+1, \ldots, k$ then the following estimates hold:

C^0-estimates:

$$(28) \quad \begin{aligned} |G_{IJ}(V,v) - H(V)_{IJ}| &\leq C H(V)_{II}^{\frac{1}{2}} H(V)_{JJ}^{\frac{1}{2}} d^2(v,P_0) \\ |G_{Ij}(V,v)| &\leq C H(V)_{II}^{\frac{1}{2}} h(v)_{jj}^{\frac{1}{2}} d^2(v,P_0) \\ |G_{ij}(V,v) - h_{ij}(v)| &\leq C h(v)_{ii}^{\frac{1}{2}} h(v)_{jj}^{\frac{1}{2}} d^2(v,P_0) \end{aligned}$$

C^1-estimates:

$$(29) \quad \begin{aligned} |\tfrac{\partial}{\partial V^I} G_{JK}(V,v)| &\leq C H(V)_{II}^{\frac{1}{2}} H(V)_{JJ}^{\frac{1}{2}} H(V)_{KK}^{\frac{1}{2}} \\ |\tfrac{\partial}{\partial v^l} G_{IJ}(V,v)| &\leq C h(v)_{ll}^{\frac{1}{2}} H(V)_{II}^{\frac{1}{2}} H(V)_{JJ}^{\frac{1}{2}} d(v,P_0) \\ |\tfrac{\partial}{\partial V^I} G_{Jj}(V,v)| &\leq C H(V)_{II}^{\frac{1}{2}} H(V)_{JJ}^{\frac{1}{2}} h(v)_{jj}^{\frac{1}{2}} d(v,P_0) \\ |\tfrac{\partial}{\partial v^l} G_{Ij}(V,v)| &\leq C H(V)_{II}^{\frac{1}{2}} h(v)_{ll}^{\frac{1}{2}} h(v)_{jj}^{\frac{1}{2}} d(v,P_0) \\ |\tfrac{\partial}{\partial V^J} G_{ij}(V,v)| &\leq C H(V)_{JJ}^{\frac{1}{2}} h(v)_{ii}^{\frac{1}{2}} h(v)_{jj}^{\frac{1}{2}} d(v,P_0) \\ |\tfrac{\partial}{\partial v^l} (G_{ij}(V,v) - h_{ij}(v))| &\leq C h(v)_{ll}^{\frac{1}{2}} h(v)_{ii}^{\frac{1}{2}} h(v)_{jj}^{\frac{1}{2}} d(v,P_0) \end{aligned}$$

C^0-estimates of the inverse:

$$(30) \quad \begin{aligned} |G^{IJ}(V,v) - H^{IJ}(V)| &\leq C H^{II}(V)^{\frac{1}{2}} H^{JJ}(V)^{\frac{1}{2}} d^2(v,P_0) \\ |G^{Ij}(V,v)| &\leq C H^{II}(V)^{\frac{1}{2}} h^{jj}(v)^{\frac{1}{2}} d^2(v,P_0) \\ |G^{ij}(V,v) - h^{ij}(v)| &\leq C h^{ii}(v)^{\frac{1}{2}} h^{jj}(v)^{\frac{1}{2}} d^2(v,P_0) \end{aligned}$$

Non-degeneracy condition for H and h with respect to the coordinates $(V^1,\ldots,V^j)$ and $(v^{j+1},\ldots,v^k)$:

$$(31) \quad \begin{aligned} H_{IJ}(V) \leq \epsilon H_{II}(V)^{\frac{1}{2}} H_{JJ}(V)^{\frac{1}{2}} (I \neq J), &\quad h_{ij}(v) \leq \epsilon h_{ii}(v)^{\frac{1}{2}} h_{jj}(v)^{\frac{1}{2}} (i \neq j) \\ H_{II}(V) H^{II}(V) \leq C, &\quad h_{ii}(v) h^{ii}(v) \leq C \end{aligned}$$

Bounds on the derivatives for H and h:

$$(32) \quad \begin{aligned} |\tfrac{\partial}{\partial V^I} H_{JK}(V)| &\leq C H_{II}(V)^{\frac{1}{2}} H_{JJ}(V)^{\frac{1}{2}} H_{KK}(V)^{\frac{1}{2}} \\ d(v,P_0)|\tfrac{\partial}{\partial v^i} h_{jk}| &\leq C h_{ii}(v)^{\frac{1}{2}} h_{jj}(v)^{\frac{1}{2}} h_{kk}(v)^{\frac{1}{2}}. \end{aligned}$$

REMARK 24. In this paper, we are interested in the case where H is the Riemannian metric $G(V,0)$, h is the Euclidean metric $h_{ij} = \delta_{ij}$ and $\tfrac{\partial}{\partial v^k} h_{ij} = \tfrac{\partial}{\partial v^l} \delta_{ij} = 0$. Thus, the above metric estimates follow immediately by Lemma 22.

REMARK 25. If G, H and h satisfy Assumption 2, then we have the following bounds for the Christoffel symbols:

$$(33) \quad \left| H_{II}^{\frac{1}{2}}{}^H\Gamma_{JK}^I \right| \leq C H_{JJ}^{\frac{1}{2}} H_{KK}^{\frac{1}{2}}, \quad \left| d(v,P_0) h_{ii}^{\frac{1}{2}}{}^h\Gamma_{jk}^i \right| \leq C h_{jj}^{\frac{1}{2}} h_{kk}^{\frac{1}{2}}.$$

Furthermore, we have the following decay estimates:
(34)
$$\left|H_{II}^{\frac{1}{2}}(\Gamma_{JK}^{I} - {}^{H}\Gamma_{JK}^{I})\right| \leq CH_{JJ}^{\frac{1}{2}}H_{KK}^{\frac{1}{2}}, \quad \left|h_{ii}^{\frac{1}{2}}(\Gamma_{jk}^{i} - {}^{h}\Gamma_{jk}^{i})\right| \leq Cd(v,P_0)h_{jj}^{\frac{1}{2}}h_{kk}^{\frac{1}{2}}$$
$$\left|H_{II}^{\frac{1}{2}}\Gamma_{Jk}^{I}\right| \leq Cd(v,P_0)H_{JJ}^{\frac{1}{2}}H_{kk}^{\frac{1}{2}}, \quad \left|H_{II}^{\frac{1}{2}}\Gamma_{jk}^{I}\right| \leq Cd(v,P_0)h_{jj}^{\frac{1}{2}}h_{kk}^{\frac{1}{2}}$$
$$\left|h_{ii}^{\frac{1}{2}}\Gamma_{jK}^{i}\right| \leq Cd(v,P_0)h_{jj}^{\frac{1}{2}}H_{KK}^{\frac{1}{2}}, \quad \left|h_{ii}^{\frac{1}{2}}\Gamma_{JK}^{i}\right| \leq Cd(v,P_0)H_{JJ}^{\frac{1}{2}}H_{KK}^{\frac{1}{2}}.$$

Indeed, Cauchy-Schwarz, (31) and (32) imply

$$\begin{aligned}
\left|d(v,P_0)h_{ii}^{\frac{1}{2}}{}^{h}\Gamma_{jk}^{i}\right| &= d(v,P_0)h_{ii}^{\frac{1}{2}}\left|h^{il}(h_{lj,k} + h_{lk,j} - h_{jk,l})\right| \\
&\leq Ch_{ii}^{\frac{1}{2}}(h^{ii}h^{ll})^{\frac{1}{2}}(h_{ll}h_{jj}h_{kk})^{\frac{1}{2}} \\
&\leq C(h_{jj}h_{kk})^{\frac{1}{2}}.
\end{aligned}$$

Furthermore, Cauchy-Schwarz, (29), (30) and (31) imply

$$\begin{aligned}
&\left|H_{II}^{\frac{1}{2}}(\Gamma_{JK}^{I} - {}^{H}\Gamma_{JK}^{I})\right| \\
&= H_{II}^{\frac{1}{2}}\left|G^{I*}(G_{*J,K} + G_{*K,J} - G_{JK,*}) - H^{I*}(H_{*J,K} + H_{*K,J} - H_{JK,*})\right| \\
&\leq H_{II}^{\frac{1}{2}}\left|(G^{IL} - H^{IL})(G_{LJ,K} + G_{LK,J} - G_{JK,L})\right| \\
&\quad + H_{II}^{\frac{1}{2}}\left|G^{Il}(G_{lJ,K} + G_{lK,J} - G_{JK,l})\right| \\
&\quad + H_{II}^{\frac{1}{2}}\left|H^{IL}(G_{LJ,K} - H_{LJ,K} + G_{LK,J} - H_{LK,J} + H_{JK,L} - G_{JK,L})\right| \\
&\leq Cd^2(v,P_0)H_{II}^{\frac{1}{2}}(H^{II}H^{LL})^{\frac{1}{2}}(H_{LL}H_{JJ}H_{KK})^{\frac{1}{2}} \\
&\quad + Cd^2(v,P_0)H_{II}^{\frac{1}{2}}(H^{II}h^{ll})^{\frac{1}{2}}(h_{ll}H_{JJ}H_{KK})^{\frac{1}{2}} \\
&\quad + CH_{II}^{\frac{1}{2}}(H^{II}H^{LL})^{\frac{1}{2}}(H_{LL}H_{JJ}H_{KK})^{\frac{1}{2}} \\
&\leq C(H_{JJ}H_{KK})^{\frac{1}{2}}.
\end{aligned}$$

The other estimates follow by similar computations.

ASSUMPTION 3. Let metrics G and h defined on $\mathbf{R}^j \times Y_2^{k-j}$ and Y_2^{k-j} satisfying Assumption 2 and
$$u = (V,v) : (B_{\sigma_*}(x_*), g) \to (\mathbf{R}^j \times Y_2^{k-j}, d_G)$$
be a harmonic map. By this, we assume that the non-singular component V of u maps into a smooth Riemannian manifold $(\mathbf{R}^j, H)$ and the singular component v of u maps into the NPC space (Y_2^{k-j}, d_h). The set $\mathcal{S}_j(u)$ satisfies the following:

(i) $v(x) = P_0$ for $x \in \mathcal{S}_j(u)$
(ii) $\dim_{\mathcal{H}}((\mathcal{S}(u) \backslash \mathcal{S}_j(u)) \cap B_{\frac{\sigma_*}{2}}(x_*)) \leq n - 2$.

REMARK 26. For a harmonic map u into a DM-complex as in (16), the fact that $v(x) = P_0$ for $x \in \mathcal{S}_j(u)$ follows from the definition of $\mathcal{S}_j(u)$. On the other hand, Assumption 3 (ii) is a part of the inductive hypothesis when we will prove Theorem 1 by a backward induction on j in Chapter 11.

ASSUMPTION 4. For $B_\sigma(x_0) \subset B_{\frac{\sigma_*}{2}}(x_*)$ and any harmonic map
$$w : (B_\sigma(x_0), g) \to (Y_2^{k-j}, h),$$

denote $\mathcal{R}(u, w)$ as the set of points x with the property that there exists a DM M of Y_2^{k-j} and $r > 0$ such that the interior of a geodesic connecting two points in $v(B_r(x)), w(B_r(x)) \subset M$. Then $\mathcal{R}(u, w)$ is of full measure in $\mathcal{R}(u) \cap B_r(x_0)$.

REMARK 27. For a harmonic map u into a DM-complex as in (16), Assumption 4 follows from Definition 17 and Corollary 19

ASSUMPTION 5. For almost every $x \in \mathcal{S}_j(u)$, we have
$$|\nabla v|^2(x) = 0 \quad \text{and} \quad |\nabla V|^2(x) = |\nabla u|^2(x).$$

REMARK 28. For a harmonic map u into a DM-complex as in (16), Assumption 5 follows by applying Lemma 29 below to Lemma 21.

By an abuse of notation, we use $|\cdot|$ to denote the norms with respect to H, h and G for maps into $\mathbf{R}^j$, Y_2^{k-j} and $\mathbf{R} \times Y_2^{k-j}$ respectively. The fact that $G(V, v)$ is asymptotically a product metric $G_0(V, v) = H(V) \oplus h(v)$ as $v \to P_0$ yields the following lemma.

LEMMA 29. Let metrics G and h defined on $\mathbf{R}^j \times Y_2^{k-j}$ and Y_2^{k-j} satisfy Assumption 2 and
$$\hat{u} : (\hat{V}, \hat{v}) : (B_{\sigma_*}(x_*), g) \to (\mathbf{R}^j \times Y_2^{k-j}, d_G)$$
be a Lipschitz map. For every $x \in \hat{\mathcal{R}}(\hat{u}) \cap B_{\sigma_*}(x_*)$ and for almost every $x \in B_{\sigma_*}(x_*)$, we have
$$\left||\nabla \hat{u}|^2(x) - \left(|\nabla \hat{V}|^2(x) + |\nabla \hat{v}|^2(x)\right)\right| \leq C d^2(\hat{v}(x), P_0)$$
where the constant C depends on the Lipschitz constant of $\hat{u}$ and the constant in the estimates (28)-(32) for the target metric G.

PROOF. We first prove that for $P, Q \in B_\lambda(P_0)$, we have
$$(35) \qquad \left(1 - C\lambda^2\right) \leq \frac{d_{H \oplus h}(P, Q)}{d_{G_\lambda}(P, Q)} \leq \left(1 + C\lambda^2\right).$$
Indeed, the properties of G imply that for any vector γ', we have
$$|\langle \gamma', \gamma' \rangle_{H \oplus h} - \langle \gamma', \gamma' \rangle_{G_\sigma}| \leq C\sigma^2 \langle \gamma', \gamma' \rangle_{H \oplus h}.$$
Let
$$\gamma : [0, d_{G_\sigma}(P, Q)] \to \mathbf{R}^{2j} \times \overline{\mathbf{H}}^{k-j}$$
be the arclength parameterized geodesic with respect to d_{G_σ} between P and Q. Then
$$\begin{aligned} d_{H \oplus h}^2(P, Q) &\leq \left(\int_0^{d_{G_\sigma}(P, Q)} \langle \gamma', \gamma' \rangle_{H \oplus h}^{\frac{1}{2}} dt\right)^2 \\ &\leq d_{G_\sigma}(P, Q) \int_0^{d_{G_\sigma}(P, Q)} \langle \gamma', \gamma' \rangle_{H \oplus h} dt \\ &\leq (1 + C\sigma^2) d_{G_\sigma}(P, Q) \int_0^{d_{G_\sigma}(P, Q)} \langle \gamma', \gamma' \rangle_G dt \\ &\leq d_{G_\sigma}^2(P, Q) \left(1 + C\sigma^2\right). \end{aligned}$$
Next, let
$$\gamma : [0, d_{H \oplus h}^2(P, Q)] \to \mathbf{R}^j \times Y_2^{k-j}$$
be the arclength parameterized geodesic with respect to $d_{H \oplus h}$ between P and Q.

Thus

$$d^2_{G_\sigma}(P,Q) \leq \left(\int_0^{d_{H\oplus h}(P,Q)} <\gamma',\gamma'>_{G_\sigma}^{\frac{1}{2}} dt\right)^2$$

$$\leq d_{H\oplus h}(P,Q) \int_0^{d_{H\oplus h}(P,Q)} <\gamma',\gamma'>_{G_\sigma} dt$$

$$\leq (1+C\sigma^2)d_{H\oplus h}(P,Q) \int_0^{d_{H\oplus h}(P,Q)} <\gamma',\gamma'>_{H\oplus h} dt$$

$$\leq d^2_{H\oplus h}(P,Q)\left(1+C\sigma^2\right).$$

This completes the proof of (35). By the definition of energy density in [**KS1**], this immediately implies for almost every $x \in B_{\sigma_*}(x_*)$ and for every $x \in B_{\sigma_*}(x_*)$ such that $\hat{u}(B_\delta(x)) \subset M$ for some DM M,

$$\left||\nabla \hat{V}|^2(x) + |\nabla \hat{v}|^2(x) - |\nabla \hat{u}|^2\right| \leq Cd^2(\hat{v}(x), P_0)$$

where C here is as in the assertion of the Lemma. □

ASSUMPTION 6. For $q \in [1,2)$ sufficiently close to 2 and any subdomain Ω compactly contained in

$$B_{\frac{\sigma_*}{2}}(x_*) \backslash \left(\mathcal{S}(u) \cap v^{-1}(P_0)\right),$$

there exists a sequence of smooth functions $\{\psi_i\}$ with $\psi_i \equiv 0$ in a neighborhood of $\mathcal{S}(u) \cap \overline{\Omega}$, $0 \leq \psi_i \leq 1$, $\psi_i \to 1$ for all $x \in \Omega \backslash \mathcal{S}(u)$ such that

$$\lim_{i \to \infty} \int_{B_{\frac{\sigma_*}{2}}(x_*)} |\nabla u||\nabla \psi_i| \, d\mu = 0,$$

$$\lim_{i \to \infty} \int_{B_{\frac{\sigma_*}{2}}(x_*)} |\nabla u||\nabla \psi_i|^q \, d\mu = 0$$

and

$$\lim_{i \to \infty} \int_{B_{\frac{\sigma_*}{2}}(x_*)} |\nabla \nabla u||\nabla \psi_i| \, d\mu = 0.$$

REMARK 30. As is the case for Assumption 3 (ii), Assumption 6 is a part of the inductive hypothesis in the proof of Theorem 1.

REMARK 31. In the sections below, we will use the following notation: Given a point $x \in \mathcal{R}(u)$, let $\mathbf{R}^j \times F$ be a DM that contains a neighborhood of $u(x) = (V(x), v(x))$. Then use the coordinates of Assumption 2 to interpret $\frac{\partial V}{\partial x^\alpha}$ as a vector in $\mathbf{R}^j$ and $\frac{\partial v}{\partial x^\alpha}$ as vectors in $\mathbf{R}^{k-j}$. For any $j \times j$-matrix $\mathcal{M}_{11}$, $j \times (k-j)$-matrix $\mathcal{M}_{12}$ and $(k-j) \times (k-j)$ matrix $\mathcal{M}_{22}$, we write

$$\mathcal{M}_{11}\nabla V \cdot \nabla V, \; \mathcal{M}_{12}\nabla V \cdot \nabla v \text{ and } \mathcal{M}_{22}\nabla v \cdot \nabla v$$

to denote the inner products defined by

$$g^{\alpha\beta}\left(\frac{\partial V}{\partial x^\alpha}\right)^T \mathcal{M}_{11}\left(\frac{\partial V}{\partial x^\beta}\right), \; g^{\alpha\beta}\left(\frac{\partial v}{\partial x^\alpha}\right)^T \mathcal{M}_{12}\left(\frac{\partial V}{\partial x^\beta}\right), \; g^{\alpha\beta}\left(\frac{\partial v}{\partial x^\alpha}\right)^T \mathcal{M}_{22}\left(\frac{\partial v}{\partial x^\beta}\right)$$

respectively. In particular, we use this notation to denote the expressions

$$\mathbf{G}_{11}(V,v)\nabla V \cdot \nabla V, \mathbf{G}_{12}(V,v)\nabla V \cdot \nabla v \text{ and } \mathbf{G}_{22}(V,v)\nabla v \cdot \nabla v$$

where we follow the notation of Lemma 22 and set
$$G = \begin{pmatrix} \mathbf{G}_{11}(V,v) & \mathbf{G}_{12}(V,v) \\ \mathbf{G}_{21}(V,v) & \mathbf{G}_{22}(V,v) \end{pmatrix}$$
with
$$\mathbf{G}_{11}(V,v) = (G_{IJ}(V,v)) \quad \mathbf{G}_{12}(V,v) = (G_{Il}(V,v))$$
$$\mathbf{G}_{21}(V,v) = (G_{lI}(V,v)) \quad \mathbf{G}_{22}(V,v) = (G_{lm}(V,v))$$
for $I, J = 1, \ldots, j$ and $l, m = j+1, \ldots, k$ to be the matrix representation of G.

CHAPTER 6

The Target Variation

The main goal of this chapter is to obtain estimates for the target variation of the singular component map $v : B_{\sigma_\star}(x_\star) \to (Y_2^{k-j}, d_h)$ of a harmonic map $u = (V, v) : B_{\sigma_\star}(x_\star) \to (\mathbf{R}^j \times Y_2^{k-j}, d_G)$ as in (16).

REMARK 32. In this section, the properties of u that we need are Assumption 2, Assumption 3 and Assumption 4 of Section 5.

Let $r_0 > 0$ such that $B_{r_0}(x_0) \subset B_{\frac{\sigma_\star}{2}}(x_\star)$ and $w : B_{r_0}(x_0) \to (Y_2^{k-j}, d_h)$ be a harmonic map. For $\sigma \in (0, r_0)$, w is Lipschitz continuous in $B_\sigma(x_0)$ by [**KS1**] Theorem 2.4.6. For $t \in [0, 1]$ and $\eta \in C_c^\infty(B_\sigma(x_0))$ with $0 \leq \eta \leq 1$, define
$$v_{t\eta} : B_\sigma(x_0) \to (Y_2^{k-j}, d_h)$$
by setting
$$v_{t\eta}(x) = (1 - t\eta(x))v(x) + t\eta(x)w(x) \tag{36}$$
where the sum indicates geometric interpolation. Furthermore, define
$$u_{t\eta} : B_\sigma(x_0) \to (\mathbf{R}^j \times Y_2^{k-j}, d_G)$$
by setting
$$u_{t\eta} = (u, v_{t\eta}). \tag{37}$$

Let $x \in B_\sigma(x_0) \cap \mathcal{R}(u, w)$; this means that there exists $\delta > 0$ and a DM $F \subset Y_2^{k-j}$ that contains $v(B_\delta(x))$ and $w(B_\delta(x))$. Since F is geodesically convex in Y_2^{k-j}, it also contains all geodesics from $v(x')$ to $w(x')$ for all $x' \in B_\delta(x)$. Hence, F contains $v_{t\eta}(x')$ for all $x' \in B_\delta(x), t \in [0, 1]$. In Lemma 33 below, we interpret $\frac{\partial v_{t\eta}}{\partial x^\beta}$ as a section of $\phi^{-1}(TF)$ where $\phi : [0,1] \times B_\delta(x) \to (Y_2^{k-j}, d_h)$ is the map $\phi(t, x) = v_{t\eta}(x)$. Furthermore, $^h\nabla$ denotes the connection on $\phi^{-1}(TF)$ induced by the Levi-Civita connection on F.

LEMMA 33. *Let $u = (V, v) : B_{\sigma_\star}(x_\star) \to (\mathbf{R}^j \times Y_2^{k-j}, d_G)$ be a harmonic map as in (16). For $v_{t\eta}$ defined in (36), there exists $C > 0$ such that for $\beta = 1, \ldots, n$ and $x \in B_\sigma(x_0) \cap \mathcal{R}(u, w)$, we have*
$$\left|{}^h\nabla_{\frac{d}{dt}} \frac{\partial v_{t\eta}}{\partial x^\beta}\right| \leq C. \tag{38}$$

PROOF. The first step is to prove the assertion under the assumption that one of the maps v or w are constant identically equal to Q_0. We will only prove the latter case since the argument for the former case is analogous. Fix $x \in B_\sigma(x_0) \cap \mathcal{R}(u, w)$ and $t \in (0, 1)$. We are also assuming $\eta \equiv 1$. Let F be a DM that contains $v(B_\delta(x))$ and Q_0 and γ be the arclength parameterized geodesic ray starting at Q_0 and ending at $v(x)$. For each $r > 0$ close to t, let $(\theta^1, \theta^2, \ldots, \theta^{k-j-1})$ be the normal coordinates centered at $\gamma(r)$ for the radius r sphere $\partial B_r(Q_0)$ in (F, h). We use this to define coordinates in a neighborhood $\mathcal{N}$ of $v_t(x)$; more specifically, the coordinates of a point P close to $v_t(x)$ is $(r, \theta^1, \ldots, \theta^{k-j-1})$ where $r = d(P, Q_0)$ and $(\theta^1, \ldots, \theta^{k-j-1})$ are the coordinates of P as a point in $\partial B_r(Q_0)$.

Since r is the distance from Q_0 and γ intersects $\partial B_r(Q_0)$ orthogonally, the components of h with respect to these coordinates satisfy

$$h_{rr} = 1, \ h_{r\theta^i} = 0 \text{ in all of } \mathcal{N}.$$

Furthermore, the choice of $(\theta^1, \ldots, \theta^{k-j-1})$ as the normal coordinates of $\partial B_r(Q_0)$ centered $\gamma(r)$ implies that

$$h_{\theta^i \theta^j} = \delta^i_j \text{ along } \gamma \text{ in } \mathcal{N}.$$

Thus, the Christoffel symbols along γ in the coordinates $(r, \theta^1, \ldots, \theta^{k-j-1})$ satisfy

$$\begin{aligned}
{}^h\Gamma^r_{rr} &= h^{rr} h_{rr,r} + h^{r\theta^i}(h_{\theta^i r, r} + h_{\theta^i r, r} - h_{rr, \theta^i}) = 0, \\
{}^h\Gamma^{\theta^k}_{rr} &= h^{\theta^k r} h_{rr,r} + h^{\theta^k \theta^i}(h_{\theta^i r, r} + h_{\theta^i r, r} - h_{rr, \theta^i}) = 0, \\
{}^h\Gamma^r_{r\theta^l} &= h^{rr}(h_{rr, \theta^l} + h_{r\theta^l, r} - h_{r\theta^l, r}) + h^{r\theta^j}(h_{\theta^j r, \theta^l} + h_{\theta^j \theta^l, r} - h_{r\theta^l, \theta^j}) = 0, \\
{}^h\Gamma^{\theta^k}_{r\theta^l} &= h^{\theta^k r}(h_{rr, \theta^l} + h_{r\theta^l, r} - h_{r\theta^l, r}) + h^{\theta^k \theta^j}(h_{\theta^j r, \theta^l} + h_{\theta^j \theta^l, r} - h_{r\theta^l, \theta^j}) = 0.
\end{aligned}$$

Using the above identities, we obtain

$$\begin{aligned}
{}^h\nabla_{\frac{d}{dt}} \frac{\partial v_t}{\partial x^\beta} &= {}^h\nabla_{\frac{d}{dt}} \frac{\partial v_t^r}{\partial x^\beta} \frac{\partial}{\partial r} + {}^h\nabla_{\frac{d}{dt}} \frac{\partial v_t^{\theta^l}}{\partial x^\beta} \frac{\partial}{\partial \theta^l} \\
&= \frac{\partial^2 v_t^r}{\partial t \partial x^\beta} \frac{\partial}{\partial r} + \frac{\partial v_t^r}{\partial x^\beta} \frac{\partial v_t^r}{\partial t} {}^h\nabla_{\frac{\partial}{\partial r}} \frac{\partial}{\partial r} + \frac{\partial v_t^{\theta^l}}{\partial x^\beta} \frac{\partial v_t^r}{\partial t} {}^h\nabla_{\frac{\partial}{\partial r}} \frac{\partial}{\partial \theta^l} \\
&= \frac{\partial^2 v_t^r}{\partial t \partial x^\beta} \frac{\partial}{\partial r} + \frac{\partial v_t^r}{\partial x^\beta} \frac{\partial v_t^r}{\partial t} \left({}^h\Gamma^r_{rr} \frac{\partial}{\partial r} + {}^h\Gamma^{\theta^k}_{rr} \frac{\partial}{\partial \theta^k} \right) \\
&\quad + \frac{\partial v_t^{\theta^l}}{\partial x^\beta} \frac{\partial v_t^r}{\partial t} \left({}^h\Gamma^r_{r\theta^l} \frac{\partial}{\partial r} + {}^h\Gamma^{\theta^k}_{r\theta^l} \frac{\partial}{\partial \theta^k} \right) \\
&= \frac{\partial^2 v_t^r}{\partial t \partial x^\beta} \frac{\partial}{\partial r} \\
&= \frac{\partial d(v, Q_0)}{\partial x^\beta} \frac{\partial}{\partial r}.
\end{aligned}$$

Thus, the assertion for this case follows with C dependent on the Lipschitz constant of v.

The second step is to consider the case when $v(x)$ and $w(x)$ are arbitrary and $\eta(x) \equiv 1$. Fix $x \in \mathcal{R}(u, w)$ and define

$$\tilde{v}_t(x') := (1-t)v(x') + tw(x)$$

and

$$\tilde{w}_t(x') := (1-t)v(x) + tw(x')$$

for x' close to x and $t \in [0, 1]$. Since $\tilde{v}_t$, $\tilde{w}_t$ and v_t are geodesic interpolation maps,

$$t \mapsto \frac{\partial \tilde{v}_t}{\partial x^\beta}(x), \ t \mapsto \frac{\partial \tilde{w}_t}{\partial x^\beta}(x) \text{ and } t \mapsto \frac{\partial v_t}{\partial x^\beta}(x)$$

are Jacobi fields along the geodesic $\gamma(t) = (1-t)v(x) + tw(x)$. Since $x \mapsto \tilde{w}_t(x)$ is constant for $t = 0$, we have

$$\frac{\partial \tilde{v}_t}{\partial x^\beta}(x)\Big|_{t=0} + \frac{\partial \tilde{w}_t}{\partial x^\beta}(x)\Big|_{t=0} = \frac{\partial v_t}{\partial x^\beta}(x)\Big|_{t=0}.$$

Similarly, since $x \mapsto \tilde{v}_t(x)$ is a constant for $t = 1$, we have

$$\frac{\partial \tilde{v}_t}{\partial x^\beta}(x)\Big|_{t=1} + \frac{\partial \tilde{w}_t}{\partial x^\beta}(x)\Big|_{t=1} = \frac{\partial v_t}{\partial x^\beta}(x)\Big|_{t=1}.$$

Thus, the uniqueness of the solution of the Jacobi equation implies that

(39) $$\frac{\partial \tilde{v}_t}{\partial x^\beta}(x) + \frac{\partial \tilde{w}_t}{\partial x^\beta}(x) = \frac{\partial v_t}{\partial x^\beta}(x), \quad \forall t \in [0,1].$$

From the first step, we obtain that

(40) $$\left| {}^h\nabla_{\frac{d}{dt}} \frac{\partial \tilde{v}_t}{\partial x^\beta}(x) \right|, \left| {}^h\nabla_{\frac{d}{dt}} \frac{\partial \tilde{w}_t}{\partial x^\beta}(x) \right| \leq C.$$

Thus, the assertion in the second step follows immediately from (39) and (40). Finally we come to the general case when η is arbitrary. If $\psi : [0,1] \times B_\delta(x) \to (Y_2^{k-j}, d_h)$ is the map $\psi(t,x) = v_t(x)$, then $\phi(t,x) = \psi(t\eta, x) = v_{t\eta}(x)$. From the second step we know that $\left| {}^h\nabla_{\frac{d}{dt}} \frac{\partial \psi(x,t)}{\partial x^\beta} \right| \leq C$, hence by the chain rule we obtain $\left| {}^h\nabla_{\frac{d}{dt}} \frac{\partial \phi(x,t)}{\partial x^\beta} \right| \leq C$. □

REMARK 34. In the case the target metric $h_{ij} = \delta_{ij}$ is Euclidean, which is the case for DM-complexes, the proof of the Lemma above is simpler. Indeed,

$$\left| \frac{d}{dt} \frac{\partial v_{t\eta}^j}{\partial x^\beta} \right| = \left| \frac{\partial}{\partial x^\beta} \eta(v^j - w^j) \right|$$
$$= \left| \eta \left(\frac{\partial v^j}{\partial x^\beta} - \frac{\partial w^j}{\partial x^\beta} \right) + \frac{\partial \eta}{\partial x^\beta}(v^j - w^j) \right| \leq C.$$

LEMMA 35. *Let* $u = (V, v) : B_{\sigma_*}(x_*) \to (\mathbf{R}^j \times Y_2^{k-j}, d_G)$ *be a harmonic map as in (16). If* $v_{t\eta}, u_{t\eta}$ *are as in (36), (37) respectively, then*

$$|\nabla u_{t\eta}|^2(x) - |\nabla u|^2(x) = |\nabla v_{t\eta}|^2(x) - |\nabla v|^2(x) + O(t^2)$$

for almost every $x \in \mathcal{S}(u)$ *where* $O(t^2)$ *is a term which is quadratic in* t.

PROOF. For $x \in \mathcal{S}_j(u)$, we have $v(x) = P_0$ by Assumption 3 (i). Thus,

$$d(v_{t\eta}(x), P_0) \leq d(v_{t\eta}(x), v(x)) + d(v(x), P_0) = t\eta d(v, w)(x).$$

Furthermore, by Lemma 29 applied with $\hat{u} = u$ and Assumption 3 (i), we have for almost every $x \in \mathcal{S}_j(u)$

$$|\nabla u|^2(x) = |\nabla V|^2(x) + |\nabla v|^2(x) + O(d^2(v, P_0)) = |\nabla V|^2(x) + |\nabla v|^2(x).$$

Finally, apply Lemma 29 with $\hat{u} = u_{t\eta}$ implies to obtain for almost every $x \in \mathcal{S}_j(u)$,

$$|\nabla u_{t\eta}|^2(x) = |\nabla V|^2(x) + |\nabla v_{t\eta}|^2(x) + O(d^2(v_{t\eta}(x), P_0))$$
$$= |\nabla V|^2(x) + |\nabla v_{t\eta}|^2(x) + O(t^2)$$

Combining the above two equations, we obtain

$$|\nabla u_{t\eta}|^2(x) - |\nabla u|^2(x) = |\nabla v_{t\eta}|^2(x) - |\nabla v|^2(x) + O(t^2), \quad \forall x \in \mathcal{S}_j(u).$$

Since $\mathcal{S}_j(u)$ is of full measure in $\mathcal{S}(u)$ by Assumption 3 (ii), this implies the assertion. □

REMARK 36. We are interested in the quantity
$$\int_{B_\sigma(x_0)} \left(|\nabla u_{t\eta}|^2 - |\nabla u|^2\right) - \left(|\nabla v_{t\eta}|^2 - |\nabla v|^2\right) d\mu.$$

We write the above integral as the sum of two terms, the first being the integral over $\mathcal{R}(u) \cap B_\sigma(x_0)$ and the second being integral over $\mathcal{S}(u) \cap B_\sigma(x_0)$. Assumption 4 implies that when we estimate the first term, we need only to estimate the integrand in the subset $\mathcal{R}(u,w)$ of $\mathcal{R}(u) \cap B_\sigma(x_0)$. Lemma 35 implies that the second term is $O(t^2)$.

The following is an estimate of the first variation for target variations.

PROPOSITION 37. Let $u = (V, v) : B_{\sigma_*}(x_*) \to (\mathbf{R}^j \times Y_2^{k-j}, d_G)$ be a harmonic map as in (16). If $w : B_{r_0}(x_0) \to (Y_2^{k-j}, d_h)$ is a harmonic map and $v_{t\eta}$, $u_{t\eta}$ are as in (36), (37) respectively, then there exists $\sigma_0 > 0$ and $C > 0$ such that
$$\limsup_{t \to 0^+} \frac{E_{x_0}^v(\sigma) - E_{x_0}^{v_{t\eta}}(\sigma)}{t} \leq C \int_{B_\sigma(x_0)} \eta d(v, P_0) d(v, w) d\mu$$
for $x_0 \in \mathcal{S}_j(u) \cap B_{\frac{\sigma_*}{2}}(x_*)$ and $\sigma \in (0, \sigma_0]$. Furthermore, C and σ_0 depend only on the constant in the estimates (28)-(32) for the target metric G, the domain metric g, the Lipschitz constants of u and w in $B_{\sigma_0}(x_0)$ and η.

PROOF. Let $x \in B_\sigma(x_0) \cap \mathcal{R}(u,w)$. Thus, there exists a DM F that contains $v_\eta(B_\delta(x))$ and $M = \mathbf{R}^j \times F$ that contains $u_\eta(B_\delta(x))$. Using coordinates of $\mathbf{R}^j \times F$, we have for $x \in B_\sigma(x_0) \cap \mathcal{R}(u,w)$ and $t_0 > 0, \tau > 0$ small

$$|\nabla u_{(t_0+\tau)\eta}|^2 - |\nabla u_{t_0\eta}|^2$$
$$= \mathbf{G}_{11}(V, v_{(t_0+\tau)\eta})\nabla V \cdot \nabla V - \mathbf{G}_{11}(V, v_{t_0\eta})\nabla V \cdot \nabla V$$
$$+ 2(\mathbf{G}_{12}(V, v_{(t_0+\tau)\eta})\nabla V \cdot \nabla v_{(t_0+\tau)\eta} - \mathbf{G}_{12}(V, v_{t_0\eta})\nabla V \cdot \nabla v_{t_0\eta})$$
$$+ \mathbf{G}_{22}(V, v_{(t_0+\tau)\eta})\nabla v_{(t_0+\tau)\eta} \cdot \nabla v_{(t_0+\tau)\eta} - \mathbf{G}_{22}(V, v_{t_0\eta})\nabla v_{t_0\eta} \cdot \nabla v_{t_0\eta}.$$

Dividing by τ, taking the limit as $\tau \to 0$, subtracting $\frac{d}{dt}\big|_{t=t_0} |\nabla v_{t\eta}|^2$ from both sides and noting that $\mathcal{R}(u,w)$ is of full measure in $\mathcal{R}(u)$ by Assumption 4, we conclude that at almost every $x \in B_\sigma(x_0) \cap \mathcal{R}(u)$ and for $t_0 > 0$ small

$$\frac{d}{dt}\bigg|_{t=t_0} \left(|\nabla u_{t\eta}|^2 - |\nabla v_{t\eta}|^2\right)$$
$$= \frac{d}{dt}\bigg|_{t=t_0} \mathbf{G}_{11}(V, v_{t\eta})\nabla V \cdot \nabla V + 2\frac{d}{dt}\bigg|_{t=t_0} \mathbf{G}_{12}(V, v_{t\eta})\nabla V \cdot \nabla v_{t\eta}$$
(41)
$$+ \frac{d}{dt}\bigg|_{t=t_0} \square(V, v_{t\eta})\nabla v_{t\eta} \cdot \nabla v_{t\eta}$$

where
$$\square(V, v) = \mathbf{G}_{22}(V, v) - h(v).$$

Since u is harmonic, we have

(42)
$$\frac{d}{dt}\bigg|_{t=0^+} \int_{B_\sigma(x_0)} |\nabla u_{t\eta}|^2 d\mu \geq 0$$

where for a function $f(t)$ defined for $t > 0$ small we set
$$\frac{d}{dt}\bigg|_{t=0^+} f := \liminf_{t \to 0^+} \frac{f(t) - f(0)}{t}.$$

By Lemma 35,
$$\int_{\mathcal{S}(u) \cap B_\sigma(x)} |\nabla u_{t\eta}|^2 - |\nabla u|^2 d\mu = \int_{\mathcal{S}(u) \cap B_\sigma(x)} |\nabla v_{t\eta}|^2 - |\nabla v|^2 d\mu + O(t^2),$$

and hence

(43) $$\frac{d}{dt}\bigg|_{t=0^+} \int_{\mathcal{S}(u) \cap B_\sigma(x)} |\nabla u_{t\eta}|^2 - |\nabla v_{t\eta}|^2 d\mu = 0.$$

Furthermore,

CLAIM 38. *For t_0 small and $C > 0$ a constant that depends only on the metric estimates (28)-(32), the Lipschitz constant of u and w and η,*
$$\left|\frac{d}{dt}\bigg|_{t=t_0} \left(|\nabla v_{t\eta}|^2 - |\nabla u_{t\eta}|^2\right)\right| \leq C, \ \forall x \in \mathcal{R}(u,w) \cap B_\sigma(x_0).$$

PROOF OF CLAIM. For $x \in \mathcal{R}(u,w)$, we use a DM to compute

(44) $$\frac{d}{dt} \mathbf{G}_{11}(V, v_{t\eta}) \nabla V \cdot \nabla V = g^{\alpha\beta} \frac{\partial}{\partial v^i} G_{IJ}(V, v_{t\eta}) \frac{dv^i_{t\eta}}{dt} \frac{\partial V^I}{\partial x^\alpha} \frac{\partial V^J}{\partial x^\beta},$$

$$\frac{d}{dt} \mathbf{G}_{12}(V, v_{t\eta}) \nabla V \cdot \nabla v_{t\eta} = g^{\alpha\beta} \frac{\partial}{\partial v^i} G_{Ij}(V, v_{t\eta}) \frac{dv^i_{t\eta}}{dt} \frac{\partial V^I}{\partial x^\alpha} \frac{\partial v^j_{t\eta}}{\partial x^\beta}$$

(45) $$+ g^{\alpha\beta} G_{Ij}(V, v_{t\eta}) \frac{\partial V^I}{\partial x^\alpha} \frac{d}{dt} \frac{\partial v^j_{t\eta}}{\partial x^\beta},$$

$$\frac{d}{dt} \Box(V, v_{t\eta}) \nabla v_{t\eta} \cdot \nabla v_{t\eta} = g^{\alpha\beta} \frac{\partial}{\partial v^i} \Box_{lj}(V, v_{t\eta}) \frac{dv^i_{t\eta}}{dt} \frac{\partial v^l_{t\eta}}{\partial x^\alpha} \frac{\partial v^j_{t\eta}}{\partial x^\beta}$$

(46) $$+ 2 g^{\alpha\beta} \Box_{lj}(V, v_{t\eta})_{lj} \frac{\partial v^l_{t\eta}}{\partial x^\alpha} \frac{d}{dt} \frac{\partial v^j_{t\eta}}{\partial x^\beta}.$$

By the Lipschitz estimate of u and (31) of Assumption 2,

(47) $$\left|H(V)^{\frac{1}{2}}_{II} \frac{\partial V^I}{\partial x^\alpha}\right|, \left|h(v)^{\frac{1}{2}}_{jj} \frac{\partial v^j}{\partial x^\alpha}\right| \leq C.$$

Since $\tau \mapsto v_{\tau\eta}(x)$ is a constant speed geodesic, we also have

(48) $$\left|h(v_{t\eta})^{\frac{1}{2}}_{jj} \frac{dv^j_{t\eta}}{dt}\right| \leq \eta d(v,w) \leq C.$$

Additionally, since
$${}^h\nabla_{\frac{d}{dt}} \frac{\partial v_{t\eta}}{\partial x^\beta} = \left(\frac{d}{dt} \frac{\partial v^i_{t\eta}}{\partial x^\beta} + \frac{\partial v^j_{t\eta}}{\partial x^\beta} \frac{\partial v^k_{t\eta}}{\partial t} {}^h\Gamma^i_{jk}\right) \frac{\partial}{\partial v^i}$$

Lemma 33, (31) and the Christoffel symbols estimates (33) imply

(49) $$d(v, P_0) \left|h^{\frac{1}{2}}_{ii} \frac{d}{dt} \frac{\partial v^i_{t\eta}}{\partial x^\beta}\right| \leq C.$$

6. THE TARGET VARIATION

Thus, the metric estimates (28), (29) along with (47), (48) and (49) imply that the absolute value of the right hand side of (44), (45) and (46) is uniformly bounded above. Combined with (41), this implies the assertion of the claim. □

We now continue with the proof of the Proposition. Since $\mathcal{R}(u,w)$ is of full measure in $\mathcal{R}(u) \cap B_\sigma(x_0)$ by Assumption 4, this immediately implies by letting t_0 go to 0

$$
\begin{aligned}
&\int_{\mathcal{R}(u)\cap B_\sigma(x_0)} \frac{d}{dt}\bigg|_{t=0^+} \left(|\nabla u_{t\eta}|^2 - |\nabla v_{t\eta}|^2\right) d\mu \\
(50) \quad &= \frac{d}{dt}\bigg|_{t=0^+} \int_{\mathcal{R}(u)\cap B_\sigma(x_0)} |\nabla u_{t\eta}|^2 - |\nabla v_{t\eta}|^2 d\mu.
\end{aligned}
$$

Therefore we conclude

$$
\begin{aligned}
&-\frac{d}{dt}\bigg|_{t=0^+} \int_{B_\sigma(x_0)} |\nabla v_{t\eta}|^2 d\mu \\
&\leq \frac{d}{dt}\bigg|_{t=0^+} \int_{B_\sigma(x_0)} |\nabla u_{t\eta}|^2 - |\nabla v_{t\eta}|^2 d\mu \quad \text{(by (42))} \\
&= \frac{d}{dt}\bigg|_{t=0^+} \int_{\mathcal{R}(u)\cap B_\sigma(x_0)} |\nabla u_{t\eta}|^2 - |\nabla v_{t\eta}|^2 d\mu \quad \text{(by (43))} \\
&= \int_{\mathcal{R}(u)\cap B_\sigma(x_0)} \frac{d}{dt}\bigg|_{t=0^+} \left(|\nabla u_{t\eta}|^2 - |\nabla v_{t\eta}|^2\right) d\mu \quad \text{(by (50))} \\
&= \int_{\mathcal{R}(u)\cap B_\sigma(x_0)} \frac{d}{dt}\bigg|_{t=0^+} \mathbf{G}_{11}(V, v_{t\eta}) \nabla V \cdot \nabla V d\mu \\
&\quad + \int_{\mathcal{R}(u)\cap B_\sigma(x_0)} \frac{d}{dt}\bigg|_{t=0^+} \mathbf{G}_{12}(V, v_{t\eta}) \nabla V \cdot \nabla v_{t\eta} d\mu \\
&\quad + \int_{\mathcal{R}(u)\cap B_\sigma(x_0)} \frac{d}{dt}\bigg|_{t=0^+} \square(V, v_{t\eta}) \nabla v_{t\eta} \cdot \nabla v_{t\eta} d\mu \quad \text{(by (41))} \\
(51) \quad &=: (I) + (II) + (III).
\end{aligned}
$$

Thus, it suffices to prove the appropriate bounds for (I), (II) and (III).

First, the metric derivative estimates (29) along with (44), (47) and (48) imply

$$
\begin{aligned}
(I) &:= \int_{\mathcal{R}(u)\cap B_\sigma(x_0)} \frac{d}{dt}\bigg|_{t=0^+} \mathbf{G}_{11}(V, v_{t\eta}) \nabla V \cdot \nabla V d\mu \\
&= \int_{\mathcal{R}(u)\cap B_\sigma(x_0)} g^{\alpha\beta} \frac{\partial}{\partial v^i} G_{IJ}(V, v) \frac{dv^i_{t\eta}}{dt}\bigg|_{t=0} \frac{\partial V^I}{\partial x^\alpha} \frac{\partial V^J}{\partial x^\beta} d\mu \\
(52) \quad &\leq C \int_{\mathcal{R}(u)\cap B_\sigma(x_0)} \eta d(v, P_0) d(v, w) d\mu.
\end{aligned}
$$

Next, by (45), we can write

$$
\begin{aligned}
(II) &:= \int_{\mathcal{R}(u) \cap B_\sigma(x_0)} \frac{d}{dt}\bigg|_{t=0^+} \mathbf{G}_{12}(V, v_{t\eta}) \nabla V \cdot \nabla v_{t\eta} d\mu \\
&= \int_{\mathcal{R}(u) \cap B_\sigma(x_0)} g^{\alpha\beta} \frac{\partial}{\partial v^i} G_{Ij}(V, v) \frac{dv^i_{t\eta}}{dt}\bigg|_{t=0} \frac{\partial V^I}{\partial x^\alpha} \frac{\partial v^j}{\partial x^\beta} d\mu \\
&\quad + \int_{\mathcal{R}(u) \cap B_\sigma(x_0)} g^{\alpha\beta} G_{Ij}(V, v) \frac{\partial V^I}{\partial x^\alpha} \frac{d}{dt}\bigg|_{t=0} \frac{\partial v^j_{t\eta}}{\partial x^\beta} d\mu
\end{aligned}
$$
(53) $\quad =: (II)_1 + (II)_2.$

As in (I),

$$
(II)_1 := \int_{\mathcal{R}(u) \cap B_\sigma(x_0)} g^{\alpha\beta} \frac{\partial}{\partial v^i} G_{Ij}(V, v) \frac{dv^i_{t\eta}}{dt}\bigg|_{t=0} \frac{\partial V^I}{\partial x^\alpha} \frac{\partial v^j}{\partial x^\beta} d\mu
$$
(54) $$ \leq C \int_{\mathcal{R}(u) \cap B_\sigma(x_0)} \eta d(v, P_0) d(v, w) d\mu. $$

Before we proceed to $(II)_2$, we will show

(55) $\quad \exists \epsilon_j \to 0 \text{ such that } \epsilon_j \mathcal{H}^{n-1}(\partial A^+_{\epsilon_j} \cap B_\sigma(x_0)) \to 0$

where

$$ A^+_\epsilon = \{x \in \overline{B_\sigma(x_0)} : d(v, P_0) > \epsilon\}. $$

Indeed, if (55) is not true, then $\epsilon \mathcal{H}^{n-1}(\partial A^+_\epsilon \cap B_\sigma(x_0)) \geq \delta > 0$ for $\epsilon < \epsilon_0$. This in turn implies

$$ \int_0^{\epsilon_0} \mathcal{H}^{n-1}(\partial A^+_\epsilon \cap B_\sigma(x_0)) d\epsilon \geq \delta \int_0^{\epsilon_0} \frac{1}{\epsilon} d\epsilon = \infty. $$

On the other hand, the co-area formula and the fact that $d(v, P_0)$ is Lipschitz imply that

$$ \int_0^\infty \mathcal{H}^{n-1}(\partial A^+_\epsilon \cap B_\sigma(x_0)) d\epsilon = \int_{A^+_0} |\nabla d(v, P_0)| d\mu < \infty. $$

This is a contradiction and this proves (55).

Let $x \in (B_\sigma(x_0) \backslash A^+_{\epsilon_j}) \cap \mathcal{R}(u, w)$. Using the metric estimate (28), we have at x

$$ |G_{Ij}(V, v)| \leq C d^2(v, P_0) H(V)^{\frac{1}{2}}_{II} h(v)^{\frac{1}{2}}_{jj}. $$

Since $\mathcal{R}(u, w)$ is of full measure in $\mathcal{R}(u)$ by Assumption 4, together with (47), (49) and the fact that $d(v, P_0) \leq \epsilon_j$ in $(\mathcal{R}(u) \cap B_\sigma(x_0)) \backslash A^+_{\epsilon_j}$ implies

$$ \int_{(\mathcal{R}(u) \cap B_\sigma(x_0)) \backslash A^+_{\epsilon_j}} g^{\alpha\beta} G_{Ij}(V, v) \frac{\partial V^I}{\partial x^\alpha} \frac{d}{dt}\bigg|_{t=0} \frac{\partial v^j_{t\eta}}{\partial x^\beta} d\mu = O(\epsilon_j), $$

and hence

(56) $$\begin{aligned}
(II)_2 &:= \int_{\mathcal{R}(u) \cap B_\sigma(x_0)} g^{\alpha\beta} G_{Ij}(V, v) \frac{\partial V^I}{\partial x^\alpha} \frac{d}{dt}\bigg|_{t=0} \frac{\partial v^j_{t\eta}}{\partial x^\beta} d\mu \\
&= \int_{A^+_{\epsilon_j}} g^{\alpha\beta} G_{Ij}(V, v) \frac{\partial V^I}{\partial x^\alpha} \frac{d}{dt}\bigg|_{t=0} \frac{\partial v^j_{t\eta}}{\partial x^\beta} d\mu + O(\epsilon_j).
\end{aligned}$$

We now apply integration by parts for the integral over $A^+_{\epsilon_j}$ above. In order to do so, let $\varrho > 0$. By [**GS**] Theorem 6.4, $\dim_\mathcal{H}(\mathcal{S}(w)) \leq n - 2$. Combined with

Assumption 3 *(ii)*, we have that $\dim_{\mathcal{H}}(\mathcal{S}(u)\backslash \mathcal{S}_j(u) \cup \mathcal{S}(w)) \leq n - 2$. Thus, there exists a cover $\{B_{r_l}(x_l) : l = 1, 2, \dots\}$ of the set $(\mathcal{S}(u)\backslash \mathcal{S}_j(u) \cup \mathcal{S}(w)) \cap A_{\epsilon_j}^+$ such that $\sum_{l=1}^{\infty} r_l^{n-1} < \varrho$. Let φ_l be a Lipschitz cut-off function which is zero in $\cup_{l=1}^{\infty} B_{r_l}(x_l)$ and identically one in $B_\sigma(x_0)\backslash \cup_{l=1}^{\infty} B_{2r_l}(x_l)$ with $|\nabla \varphi_l| \leq 2r_l^{-1}$ in $B_{r_l}(x_l)$. Thus, with $\varphi_\varrho = \Pi_l^\infty \varphi_l$, we have

$$\int_{A_{\epsilon_j}^+} g^{\alpha\beta} G_{Ij}(V,v) \frac{\partial V^I}{\partial x^\alpha} \frac{d}{dt}\bigg|_{t=0} \frac{\partial v_{t\eta}^j}{\partial x^\beta} d\mu$$

$$= \lim_{\varrho \to 0} \int_{A_{\epsilon_j}^+} \varphi_\varrho g^{\alpha\beta} G_{Ij}(V,v) \frac{\partial V^I}{\partial x^\alpha} \frac{d}{dt}\bigg|_{t=0} \frac{\partial v_{t\eta}^j}{\partial x^\beta} d\mu$$

$$= \lim_{\varrho \to 0} \left[-\int_{A_{\epsilon_j}^+} \varphi_\varrho \frac{1}{\sqrt{g}} \frac{\partial}{\partial x^\beta}(\sqrt{g} g^{\alpha\beta} \frac{\partial V^I}{\partial x^\alpha}) G_{Ij}(V,v) \frac{dv_{t\eta}^j}{dt}\bigg|_{t=0} d\mu \right.$$

$$- \int_{A_{\epsilon_j}^+} \varphi_\varrho g^{\alpha\beta} \frac{\partial}{\partial x^\beta} G_{Ij}(V,v) \frac{\partial V^I}{\partial x^\alpha} \frac{dv_{t\eta}^j}{dt}\bigg|_{t=0} d\mu$$

$$- \int_{A_{\epsilon_j}^+} g^{\alpha\beta} G_{Ij}(V,v) \frac{\partial V^I}{\partial x^\alpha} \frac{\partial \varphi_\varrho}{\partial x^\beta} \frac{dv_{t\eta}^j}{dt}\bigg|_{t=0} d\mu$$

$$+ \left. \int_{\partial A_{\epsilon_j}^+} \varphi_\varrho g^{\alpha\beta} G_{Ij}(V,v) \frac{\partial V^I}{\partial x^\alpha} \frac{dv_{t\eta}^j}{dt}\bigg|_{t=0} \left(\vec{n} \cdot \frac{\partial}{\partial x^\beta}\right) d\Sigma \right]$$

(57) $$=: \lim_{\varrho \to 0} [(II)_{21} + (II)_{22} + (II)_{23} + (II)_{24}].$$

As a component function of a harmonic map u, V^I satisfies the equation

$$\frac{1}{\sqrt{g}} \frac{\partial}{\partial x^\beta}\left(\sqrt{g} g^{\alpha\beta} \frac{\partial V^I}{\partial x^\alpha}\right) \frac{\partial}{\partial V^I}$$

$$= -g^{\alpha\beta} \left(\Gamma^I_{JK}(V,v) \frac{\partial V^J}{\partial x^\alpha} \frac{\partial V^K}{\partial x^\beta} + \Gamma^I_{Ji}(V,v) \frac{\partial V^J}{\partial x^\alpha} \frac{\partial v^i}{\partial x^\beta} + \Gamma^I_{ij}(V,v) \frac{\partial v^i}{\partial x^\alpha} \frac{\partial v^j}{\partial x^\beta} \right) \frac{\partial}{\partial V^I}$$

in a neighborhood of a regular point $x \in A_\epsilon^+ \cap \mathcal{R}(u)$. By the Christoffel symbols estimates (33), (34) and the Lipschitz estimates (47), we obtain

(58) $$\left| \frac{1}{\sqrt{g}} \frac{\partial}{\partial x^\beta}\left(\sqrt{g} g^{\alpha\beta} \frac{\partial V^I}{\partial x^\alpha}\right) H_{II}^{\frac{1}{2}} \right| \leq C.$$

Thus, the metric estimates (28) and (48) imply

$$(II)_{21} := -\int_{A_{\epsilon_j}^+} \varphi_\varrho \frac{1}{\sqrt{g}} \frac{\partial}{\partial x^\beta}(\sqrt{g} g^{\alpha\beta} \frac{\partial V^I}{\partial x^\alpha}) G_{Ij}(V,v) \frac{dv_{t\eta}^j}{dt}\bigg|_{t=0} d\mu$$

$$\leq C \int_{B_\sigma(x_0)} \eta d^2(v,P_0) d(v,w) d\mu$$

(59) $$\leq C \int_{B_\sigma(x_0)} \eta d(v,P_0) d(v,w) d\mu.$$

6. THE TARGET VARIATION

By the metric derivative estimates (29) and the Lipschitz estimates (47) we obtain

$$\left|\frac{\partial}{\partial x^\beta}G_{Ij}(V,v)\right| = \left|\frac{\partial}{\partial V^J}G_{Ij}(V,v)\frac{\partial V^J}{\partial x^\beta} + \frac{\partial}{\partial v^k}G_{Ij}(V,v)\frac{\partial v^k}{\partial x^\beta}\right|$$

(60)
$$\leq Cd(v,P_0)H(V)_{II}^{\frac{1}{2}}h(v)_{jj}^{\frac{1}{2}}.$$

Combined with (47) and (48), this implies

$$(II)_{22} := -\int_{A_{\epsilon_j}^+}\varphi_\varrho g^{\alpha\beta}\frac{\partial}{\partial x^\beta}G_{Ij}(V,v)\frac{\partial V^I}{\partial x^\alpha}\frac{dv_{t\eta}^j}{dt}\bigg|_{t=0}d\mu$$

(61)
$$\leq C\int_{B_\sigma(x_0)}\eta d(v,P_0)d(v,w)d\mu.$$

By the properties of the set of cut-off functions $\{\varphi_l\}$, we have

$$(II)_{23} := -\int_{A_{\epsilon_j}^+}g^{\alpha\beta}G_{Ij}(V,v)\frac{\partial V^I}{\partial x^\alpha}\frac{\partial \varphi_\varrho}{\partial x^\beta}\frac{dv_{t\eta}^j}{dt}\bigg|_{t=0}d\mu$$

$$\leq C\sum_{l=1}^L\int_{B_{r_l}(x_l)}|\nabla\varphi_l|d\mu$$

$$\leq C\sum_{l=1}^L\frac{1}{r_l}\mathrm{Vol}(B_{r_l}(x_l))$$

(62)
$$\leq C\sum_{l=1}^L r_l^{n-1} = O(\varrho).$$

Furthermore, $|G_{Ii}(V,v)| \leq C\epsilon_j^2 H(V)_{II}^{\frac{1}{2}}h(v)_{ii}^{\frac{1}{2}}$ on $\partial A_{\epsilon_j}^+$ by the metric estimates (28), and hence

$$\left|\int_{\partial A_{\epsilon_j}^+ \cap B_\sigma(x_0)}\varphi_\varrho g^{\alpha\beta}G_{Ij}(V,v)\frac{\partial V^I}{\partial x^\alpha}\frac{dv_{t\eta}^j}{dt}\bigg|_{t=0}\left(\vec{n}\cdot\frac{\partial}{\partial x^\beta}\right)d\Sigma\right|$$

$$\leq C\epsilon_j^2 \mathcal{H}^{n-1}(\partial A_{\epsilon_j}^+ \cap B_\sigma(x_0)) = O(\epsilon_j)$$

where we have used (55) for the last equality. Lastly, the fact that η has compact support in $B_\sigma(x_0)$ implies $\frac{dv_{t\eta}^j}{dt}\big|_{t=0} = 0$ on $\partial B_\sigma(x_0)$. Thus,

$$\int_{A_{\epsilon_j}^+ \cap \partial B_\sigma(x_0)}\varphi_\varrho g^{\alpha\beta}G_{Ij}(V,v)\frac{\partial V^I}{\partial x^\alpha}\frac{dv_{t\eta}^j}{dt}\bigg|_{t=0}\left(\vec{n}\cdot\frac{\partial}{\partial x^\beta}\right)d\Sigma = 0.$$

The above two inequalities imply

$$(II)_{24} := \int_{\partial A_{\epsilon_j}^+}\varphi_\varrho g^{\alpha\beta}G_{Ij}(V,v)\frac{\partial V^I}{\partial x^\alpha}\frac{dv_{t\eta}^j}{dt}\bigg|_{t=0}\left(\vec{n}\cdot\frac{\partial}{\partial x^\beta}\right)d\Sigma$$

$$= \int_{A_{\epsilon_j}^+ \cap \partial B_\sigma(x_0)} + \int_{\partial A_{\epsilon_j}^+ \cap B_\sigma(x_0)}\varphi_\varrho g^{\alpha\beta}G_{Ij}(V,v)\frac{\partial V^I}{\partial x^\alpha}\frac{dv_{t\eta}^j}{dt}\bigg|_{t=0}\left(\vec{n}\cdot\frac{\partial}{\partial x^\beta}\right)d\Sigma$$

(63)
$$= O(\epsilon_j).$$

Combining (56), (57), (59), (61), (62), (63), and letting $\epsilon_j, \varrho \to 0$, we obtain

$$(64) \quad (II)_2 \leq C \int_{B_\sigma(x_0)} \eta d(v, P_0) d(v, w) \, d\mu.$$

Combining (53), (54), (64), we have

$$(II) := \int_{\mathcal{R}(u) \cap B_\sigma(x_0)} \frac{d}{dt}\bigg|_{t=0^+} \mathbf{G}_{12}(V, v_{t\eta}) \nabla V \cdot \nabla v_{t\eta} d\mu$$

$$(65) \quad \leq C \int_{B_\sigma(x_0)} \eta d(v, P_0) d(v, w) \, d\mu.$$

Finally, by (46), we can write

$$(III) = \int_{\mathcal{R}(u) \cap B_\sigma(x_0)} \frac{d}{dt}\bigg|_{t=0^+} \Box(V, v_{t\eta}) \nabla v_{t\eta} \cdot \nabla v_{t\eta} d\mu$$

$$= \int_{\mathcal{R}(u) \cap B_\sigma(x_0)} g^{\alpha\beta} \frac{\partial}{\partial v^i} \Box_{lj}(V, v) \frac{dv^i_{t\eta}}{dt}\bigg|_{t=0} \frac{\partial v^l}{\partial x^\alpha} \frac{\partial v^j}{\partial x^\beta} d\mu$$

$$+ \int_{\mathcal{R}(u) \cap B_\sigma(x_0)} g^{\alpha\beta} \Box_{lj}(V, v) \frac{\partial v^l}{\partial x^\alpha} \frac{d}{dt}\bigg|_{t=0} \frac{\partial v^j_{t\eta}}{\partial x^\beta} d\mu$$

$$(66) \quad =: (III)_1 + (III)_2.$$

We derive an estimate for $(III)_1$ in exactly the same way as in (I) noting that similarity of the the metric derivative estimatess (29) for $\mathbf{G}_{11}(V, v)$ and $\Box(V, v)$. We obtain

$$(III)_1 := \int_{\mathcal{R}(u) \cap B_\sigma(x_0)} g^{\alpha\beta} \frac{\partial}{\partial v^l} \Box_{ij}(V, v) \frac{dv^l_{t\eta}}{dt}\bigg|_{t=0} \frac{\partial v^i}{\partial x^\alpha} \frac{\partial v^j}{\partial x^\beta} d\mu$$

$$(67) \quad \leq C \int_{\mathcal{R}(u) \cap B_\sigma(x_0)} \eta d(v, P_0) d(v, w) d\mu.$$

To estimate $(III)_2$, we write similarly to $(II)_2$

$$(III)_2 = \int_{A^+_{\epsilon_j}} g^{\alpha\beta} \Box_{ij}(V, v) \frac{\partial v^i}{\partial x^\alpha} \frac{d}{dt}\bigg|_{t=0} \frac{\partial v^j_{t\eta}}{\partial x^\beta} d\mu + O(\epsilon_j)$$

$$= \lim_{\varrho \to 0} \int_{A^+_{\epsilon_j}} \varphi_\varrho g^{\alpha\beta} \Box_{ij}(V, v) \frac{\partial v^i}{\partial x^\alpha} \frac{d}{dt}\bigg|_{t=0} \frac{\partial v^j_{t\eta}}{\partial x^\beta} d\mu + O(\epsilon_j)$$

$$= \lim_{\varrho \to 0} \bigg[- \int_{A^+_{\epsilon_j}} \varphi_\varrho \frac{1}{\sqrt{g}} \frac{\partial}{\partial x^\beta}(\sqrt{g} g^{\alpha\beta} \frac{\partial v^i}{\partial x^\alpha}) \Box_{ij}(V, v) \frac{dv^j_{t\eta}}{dt}\bigg|_{t=0} d\mu$$

$$- \int_{A^+_{\epsilon_j}} \varphi_\varrho g^{\alpha\beta} \frac{\partial}{\partial x^\beta} \Box_{ij}(V, v) \frac{\partial v^i}{\partial x^\alpha} \frac{dv^j_{t\eta}}{dt}\bigg|_{t=0} d\mu$$

$$- \int_{A^+_{\epsilon_j}} g^{\alpha\beta} \Box_{ij}(V, v) \frac{\partial v^i}{\partial x^\alpha} \frac{\partial \varphi_\varrho}{\partial x^\beta} \frac{dv^j_{t\eta}}{dt}\bigg|_{t=0} d\mu$$

$$+ \int_{\partial A^+_{\epsilon_j}} \varphi_\varrho g^{\alpha\beta} \Box_{ij}(V, v) \frac{\partial v^i}{\partial x^\alpha} \frac{dv^j_{t\eta}}{dt}\bigg|_{t=0} \left(\vec{n} \cdot \frac{\partial}{\partial x^\beta}\right) d\Sigma \bigg] + O(\epsilon_j)$$

$$=: \lim_{\varrho \to 0}[(III)_{21} + (III)_{22} + (III)_{23} + (III)_{24}] + O(\epsilon_j).$$

We obtain the estimates for $(III)_{22}$, $(III)_{23}$ and $(III)_{24}$ in exactly the same way as for $(II)_{22}$, $(II)_{23}$ and $(II)_{24}$ after noting the similarity of the the metric estimates (28) and (29) for $\mathbf{G}_{12}(V,v)$ and $\square(V,v) = \mathbf{G}_{22}(V,v) - h(v)$. Furthermore, we obtain the estimates for $(III)_{21}$ analogously to $(II)_{21}$. Indeed, as a component function of a harmonic map u, v^i satisfies the equation

$$\frac{1}{\sqrt{g}} \frac{\partial}{\partial x^\beta} \left(\sqrt{g} g^{\alpha\beta} \frac{\partial v^i}{\partial x^\alpha} \right) \frac{\partial}{\partial v^i}$$
$$= -g^{\alpha\beta} \left(\Gamma^i_{JK}(V,v) \frac{\partial V^J}{\partial x^\alpha} \frac{\partial V^K}{\partial x^\beta} + \Gamma^i_{Ji}(V,v) \frac{\partial V^J}{\partial x^\alpha} \frac{\partial v^i}{\partial x^\beta} + \Gamma^i_{jk}(V,v) \frac{\partial v^j}{\partial x^\alpha} \frac{\partial v^k}{\partial x^\beta} \right) \frac{\partial}{\partial v^i}$$

in a neighborhood of a regular point $x \in A_\epsilon^+ \cap \mathcal{R}(u)$. By the Christoffel symbols estimates (33), (34) and the Lipschitz estimates (47), we obtain

$$(68) \qquad d(v, P_0) \left| \frac{1}{\sqrt{g}} \frac{\partial}{\partial x^\beta} \left(\sqrt{g} g^{\alpha\beta} \frac{\partial v^i}{\partial x^\alpha} \right) h_{ii}^{\frac{1}{2}} \right| \leq C.$$

Hence,

$$(III) := \int_{\mathcal{R}(u) \cap B_\sigma(x_0)} \frac{d}{dt}\bigg|_{t=0^+} \square(V, v_{t\eta}) \nabla v_{t\eta} \cdot \nabla v_{t\eta} \, d\mu$$
$$(69) \qquad \leq C \int_{B_\sigma(x_0)} \eta d(v, P_0) d(v, w) \, d\mu.$$

Thus, the assertion of the lemma follows from (51), (52), (65) and (69). □

With w equal to the constant map P_0 in Proposition 37, we obtain

COROLLARY 39. *Let $u = (V,v) : B_{\sigma_*}(x_*) \to (\mathbf{R}^j \times Y_2^{k-j}, d_G)$ be a harmonic map as in (16). There exists $\sigma_0 > 0$ and $C > 0$ such that*

$$-C \int_{B_\sigma(x_0)} \eta d^2(v, P_0) d\mu + 2 \int_{B_\sigma(x_0)} \eta |\nabla v|^2 d\mu < - \int_{B_\sigma(x_0)} \nabla \eta \cdot \nabla d^2(v, P_0) \, d\mu$$

for $x_0 \in \mathcal{S}_j(u) \cap B_{\frac{\sigma_*}{2}}(x_*)$, $\sigma \in (0, \sigma_0]$ and $\eta \in C_c^\infty(B_\sigma(x_0))$ with $0 \leq \eta \leq 1$. Furthermore, C and σ_0 depend only on the constant in the estimates (28)-(32) for the target metric G, the domain metric g and the Lipschitz constants of u and w in $B_{\sigma_0}(x_0)$.

PROOF. From [GS] Section 2,

$$E_{x_0}^{v_{t\eta}}(\sigma) \leq \int_{B_\sigma(x_0)} (1 - t\eta)^2 |\nabla v|^2 d\mu - t \int_{B_\sigma(x_0)} \nabla \eta \cdot \nabla d^2(v(x), P_0) d\mu + 0(t^2).$$

Hence rearranging terms, dividing by t and letting $t \to 0$, we obtain

$$(70) \qquad 2 \int_{B_\sigma(x_0)} \eta |\nabla v|^2 d\mu$$
$$\leq - \int_{B_\sigma(x_0)} \nabla \eta \cdot \nabla d^2(v(x), P_0) \, d\mu + \liminf_{t \to 0^+} \frac{E^v(\sigma) - E^{v_{t\eta}}(\sigma)}{t}.$$

Proposition 37 with $w = P_0$ implies

$$\liminf_{t \to 0^+} \frac{E_{x_0}^v(\sigma) - E_{x_0}^{v_{t\eta}}(\sigma)}{t} \leq C \int_{B_\sigma(x_0)} \eta d^2(v, P_0) d\mu.$$

Combining the above two, we obtain the assertion of the Proposition. □

The following is the analogue of the target variation formula in [**GS**].

PROPOSITION 40. *Let $u = (V, v) : B_{\sigma_*}(x_*) \to (\mathbf{R}^j \times Y_2^{k-j}, d_G)$ be a harmonic map as in (16). There exists $C > 0$ such that for $\sigma \in (0, \sigma_0)$,*

$$2E_{x_0}^v(\sigma) \leq \int_{\partial B_\sigma(x_0)} \frac{\partial}{\partial r} d^2(v, P_0) d\Sigma + C \int_{B_\sigma(x_0)} d^2(v, P_0) d\mu. \tag{71}$$

Furthermore, C and σ_0 depends only on the constant in the estimates (28)-(32) for the target metric G, the domain metric g and the Lipschitz constants of u and w in $B_{\sigma_0}(x_0)$.

PROOF. Follows immediately from letting η approximate the characteristic function of $B_\sigma(x_0)$ in Corollary 39. □

REMARK 41. When (71) is compared with [**GS**] inequality (2.2), we note the additional error term of $C \int_{B_\sigma(x_0)} d^2(v, P_0) d\mu$. Furthermore, Corollary 39 says that the function $d^2(v, P_0)$ is *almost subharmonic* up to the same error term.

CHAPTER 7

Lower Order Bound

The main goal of this section is to prove the Poincare type inequality (cf. Lemma 43).

REMARK 42. In this section, the properties of u that we need are Assumption 1, Assumption 2, Assumption 3 and Assumption 4 of Section 5.

PROPOSITION 43. *Let $u = (V,v) : B_{\sigma_\star}(x_\star) \to (\mathbf{R}^j \times Y_2^{k-j}, d_G)$ be a harmonic map as in (16). Then for any $\epsilon_0 > 0$, there exists $R_0 > 0$ such that*

$$(72) \qquad 1 - \epsilon_0 \leq \frac{\sigma E_{x_i}^v(\sigma)}{I_{x_i}^v(\sigma)}, \forall x_i \in \mathcal{S}_j(u) \cap B_{\frac{\sigma_\star}{2}}(x_\star), \ \sigma \in (0, R_0).$$

Before we proceed with the proof of Proposition 43, we need some preliminary material. Let $u = (V,v) : B_{\sigma_\star}(x_\star) \to (\mathbf{R}^j \times Y_2^{k-j}, d_G)$ as in Proposition 43, $x \in \mathcal{S}_j(u)$ and $\sigma > 0$ sufficiently small such that $B_\sigma(x) \subset B_{\frac{\sigma_\star}{2}}(x_\star)$. Note that $x \in \mathcal{S}_j(u)$ implies $v(x) = P_0$ (cf. Assumption 3 (i)). Use normal coordinates to identify the σ-ball about x with $(B_\sigma(0), g_x)$ where $B_\sigma(0) \subset \mathbf{R}^n$. We define the restriction maps

$$_{x,\sigma}v : (B_\sigma(0), g_x) \to Y_2^{k-j}, \ _{\sigma,x}v = v\big|_{B_\sigma(0)},$$

the harmonic maps

$$_{x,\sigma}w : (B_\sigma(0), g_x) \to (Y_2^{k-j}, d) \text{ with } \ _{\sigma,x}w\big|_{\partial B_\sigma(0)} = \ _{\sigma,x}v\big|_{\partial B_\sigma(0)}$$

and set

$$(73) \qquad \nu_{\sigma,x} = \left(\frac{I_0^{\sigma,x v}(\sigma)}{\sigma^{n-1}}\right)^{1/2}.$$

Let $g_{\sigma,x}(y) = g_x(\sigma y)$ be the rescaled metric on $B_1(0)$. Using the homogeneous structure of Y_2^{k-j} (cf. Assumption 1), define the rescaled maps

$$v_{\sigma,x}, w_{\sigma,x} : (B_1(0), g_{\sigma,x}) \to (Y_2^{k-j}, d)$$

by setting

$$v_{\sigma,x}(y) = \nu_{\sigma,x}^{-1} \ _{\sigma,x}v(\sigma y) \text{ and } w_{\sigma,x}(y) = \nu_{\sigma,x}^{-1} \ _{\sigma,x}w(\sigma y).$$

We will denote by $d\mu_{\sigma,x}$, $d\Sigma_{\sigma,x}$ the volume forms on $B_1(0)$, $\partial B_r(0)$ respectively with respect to the metric $g_{\sigma,x}$. The normalization by $\nu_{\sigma,x}$ implies that

$$I_0^{v_{\sigma,x}}(1) = 1.$$

DEFINITION 44. *The maps $\{v_{\sigma,x}\}_{\sigma>0}$ are called the blow-up maps of v at x and the maps $\{w_{\sigma,x}\}$ are called the approximating harmonic blow-up maps of v at x.* We will drop the subscript x from $v_{\sigma,x}, w_{\sigma,x}, \ _{\sigma,x}v, \ _{\sigma,x}w, \ g_{\sigma,x}, \ d\mu_{\sigma,x}$ and $d\Sigma_{\sigma,x}$ above when it is clear at which point we are taking the blow ups. Note that in this notation v_σ may be different from the second component $\pi_2 \circ u_\sigma$ of u_σ as the

37

blow-up factors μ, ν for u, v respectively may be different. Hopefully, this will not cause any confusion to the reader since it will be clear from the context which one we are using. Furthermore, we will drop the subscript x from E_x and I_x when the point is understood.

LEMMA 45. *Let $u = (V, v) : B_{\sigma_*}(x_*) \to (\mathbf{R}^j \times Y_2^{k-j}, d_G)$ be a harmonic map as in (16) and ${}_\sigma v$, ${}_\sigma w$, v_σ, w_σ as in Definition 44. Then there exists a constant $C > 0$ depending only on the domain metric g such that*

(74) $$\int_{B_\sigma(0)} d^2(v, {}_\sigma w) d\mu \leq C\sigma^2 \left(E^v(\sigma) - E^{\sigma w}(\sigma) \right)$$

$$\int_{B_1(0)} d^2(v_\sigma, w_\sigma) d\mu_\sigma \leq C \left(E^{v_\sigma}(1) - E^{w_\sigma}(1) \right)$$

(75) $$\int_{B_\sigma(0)} |\nabla d(v, {}_\sigma w)|^2 d\mu \leq E^v(\sigma) - E^{\sigma w}(\sigma)$$

$$\int_{B_1(0)} |\nabla d(v_\sigma, w_\sigma)|^2 d\mu_\sigma \leq E^{v_\sigma}(1) - E^{w_\sigma}(1)$$

(76) $$\int_{B_\sigma(0)} d^2({}_\sigma w, P_0) \, d\mu \leq C\sigma I^v(\sigma)$$

$$\int_{B_1(0)} d^2(w_\sigma, P_0) \, d\mu_\sigma \leq C$$

(77) $$\int_{B_\sigma(0)} d^2(v, P_0) \, d\mu \leq C \left(\sigma I^v(\sigma) + \sigma^2 E^v(\sigma) \right)$$

$$\int_{B_1(0)} d^2(v_\sigma, P_0) \, d\mu_\sigma \leq C \left(1 + E^{v_\sigma}(1) \right).$$

PROOF. It will be sufficient to prove (74), (75), (76) and (77) since the other inequalities will then follow after a change of variables $x = \sigma y$ and a multiplication by ν_σ^{-2}. Let ${}_\sigma w_{\frac{1}{2}} : B_\sigma \to (Y_2^{k-j}, d_h)$ be the map defined by setting ${}_\sigma w_{\frac{1}{2}}(x)$ to be the midpoint of the geodesic between $v(x)$ and ${}_\sigma w(x)$. Then by (2.2iv) of [**KS2**], we have

$$2 \, E^{\sigma w_{\frac{1}{2}}}(\sigma) \leq E^v(\sigma) + E^{\sigma w}(\sigma) - \int_{B_\sigma(0)} |\nabla d(v, {}_\sigma w)|^2 d\mu.$$

The harmonicity of ${}_\sigma w$ implies $E^{\sigma w}(\sigma) \leq E^{\sigma w_{\frac{1}{2}}}(\sigma)$ which in turn implies (75). Let $C > 0$ be a generic constant depending only on the domain metric g. The Poincare inequality then implies that

$$\int_{B_\sigma(0)} d^2(v, {}_\sigma w) d\mu \leq C\sigma^2 \int_{B_\sigma(0)} |\nabla d(v, {}_\sigma w)|^2 d\mu.$$

Combining the above two inequality, we obtain (74). Since ${}_\sigma w$ is a harmonic map (cf. [**GS**], last formula on p. 195),

$$I^{\sigma w}(s) \leq e^{C\sigma^2} \frac{I^{\sigma w}(\sigma)}{\sigma^{n+1}} s^{n+1}, \quad \text{for } s \leq \sigma.$$

Integrating over $s \in (0, \sigma)$, there exists a constant $C > 0$ depending only on g such that
$$\int_{B_\sigma(0)} d^2({_\sigma}w, P_0) \, d\mu \leq C\sigma \int_{\partial B_\sigma(0)} d^2({_\sigma}w, P_0) \, d\Sigma = C\sigma \, I^v(\sigma)$$
which proves (76). The inequality (77) follows immediately from the triangle inequality and (74). □

LEMMA 46. *Let $u = (V, v) : B_{\sigma_*}(x_*) \to (\mathbf{R}^j \times Y_2^{k-j}, d_G)$ be a harmonic map as in (16), v_σ, w_σ be as in Definition 44 and assume there exists $A > 0$ such that $E^{v_\sigma}(1) \leq A$. Then there exists a constant $C > 0$ such that*
$$E^{v_\sigma}(1) - E^{w_\sigma}(1) \leq C\sigma^2.$$

Furthermore, C depends only on the constant in the estimates (28)-(32) for the target metric G, the domain metric g, the Lipschitz constant of u and A.

PROOF. Let $\hat{u} = (V, {_\sigma}w)$. By Lemma 29,
$$|\nabla v|^2 \leq |\nabla u|^2 - |\nabla V|^2 + Cd^2(v, P_0)$$
$$-|\nabla_\sigma w|^2 \leq -|\nabla \hat{u}|^2 + |\nabla V|^2 + Cd^2({_\sigma}w, P_0),$$
and thus
(78) $\quad |\nabla v|^2 - |\nabla_\sigma w|^2 \leq |\nabla u|^2 - |\nabla \hat{u}|^2 + Cd^2(v, P_0) + Cd^2({_\sigma}w, P_0).$

Integrating over $B_\sigma(x_0)$, we obtain
$$E^v(\sigma) - E^{{_\sigma}w}(\sigma) \leq E^u(\sigma) - E^{\hat{u}}(\sigma) + C \int_{B_\sigma(x_0)} d^2(v, P_0) + d^2({_\sigma}w, P_0) d\mu.$$

Harmonicity of u and scaling immediately implies
$$E^{v_\sigma}(1) - E^{w_\sigma}(1) \leq C\sigma^2 \int_{B_1(0)} d^2(v_\sigma, P_0) + d^2(w_\sigma, P_0) d\mu_\sigma$$
$$\leq C\sigma^2(1 + E^{v_\sigma}(1)) \text{ (by Lemma 45)}$$
where $d\mu_\sigma$ is the volume form with respect to metric g_σ. Since $E^{v_\sigma}(1) \leq A$, the proof is complete. □

The following states that $d^2(v_\sigma, w_\sigma)$ is close to being a subharmonic function.

LEMMA 47. *Let $u = (V, v) : B_{\sigma_*}(x_*) \to (\mathbf{R}^j \times Y_2^{k-j}, d_G)$ be a harmonic map as in (16), $x \in \mathcal{S}_j(u) \cap B_{\frac{\sigma_*}{2}}(x_*)$, identify $x = 0$ via normal coordinates and let $\sigma > 0$ sufficiently small such that $B_\sigma(0) \subset B_{\frac{\sigma_*}{2}}(x_*)$. For v_σ and g_σ as in Definition 44, assume there exists $A > 0$ such that $E^{v_\sigma}(1) \leq A$. Then for $\rho \in [\frac{3}{4}, 1]$, a harmonic map $w : (B_\rho(0), g_\sigma) \to Y_2^{k-j}$ with $E^w(\rho) \leq A$ and $\sigma_0 \in (0, 1)$, there exists a constant $C > 0$ such that if $\eta \in C_c(B_{\rho\sigma_0}(0))$, then*
$$-C\sigma^2 \int_{B_\rho(0)} \eta d(v_\sigma, P_0) d(v_\sigma, w) d\mu_\sigma \leq -\int_{B_\rho(0)} \nabla \eta \cdot \nabla d^2(v_\sigma, w) \, d\mu_\sigma.$$

Furthermore, C depends only on σ_0, the constant in the estimates (28)-(32) for the target metric G, the domain metric g, the Lipschitz constant of u and A.

PROOF. We apply Proposition 37 with w replaced by the harmonic map $\hat{w} : B_{\rho\sigma}(0) \to Y_2^{k-j}$ defined by setting $\hat{w}(y) = \nu_\sigma w(\frac{y}{\sigma})$, σ_0 replaced by $\rho\sigma$, σ replaced by $\rho\sigma\sigma_0$ and x_0 replaced by x. Thus, for a non-negative smooth function $\eta \in C_c^\infty(B_{\rho\sigma\sigma_0}(x))$ with $0 \leq \eta \leq 1$ and t sufficiently small,

$$(79) \quad \limsup_{t \to 0^+} \frac{E_0^v(\rho\sigma\sigma_0) - E_0^{v_{t\eta}}(\rho\sigma\sigma_0)}{t} \leq C \int_{B_{\rho\sigma\sigma_0}(x)} \eta d(v, P_0) d(v, \hat{w}) d\mu$$

where $C > 0$ depends only on σ_0, the constant in the estimates (28)-(32) for the target metric G, the domain metric g and the Lipschitz constants of u and $\hat{w}$ in $B_{\rho\sigma\sigma_0}(x)$. Using the fact that $\hat{w}$ is energy minimizing, we obtain from [**KS1**] Lemma 2.4.2 that

$$E_0^{v_{t\eta}}(\rho\sigma\sigma_0) \leq E_0^v(\rho\sigma\sigma_0) - t \int_{B_{\rho\sigma\sigma_0}(x)} \nabla\eta \cdot \nabla d^2(v, \hat{w}) d\mu + 0(t^2).$$

Hence rearranging terms, dividing by t and letting $t \to 0$, we obtain

$$0 \leq -\int_{B_{\rho\sigma\sigma_0}(x)} \nabla\eta \cdot \nabla d^2(v, \hat{w}) \, d\mu + \limsup_{t \to 0^+} \frac{E_0^v(\rho\sigma\sigma_0) - E_0^{v_{t\eta}}(\rho\sigma\sigma_0)}{t}.$$

Combining this with (79), we obtain

$$-C \int_{B_{\rho\sigma\sigma_0}(x)} \eta d(v, P_0) d(v, \hat{w}) d\mu \leq -\int_{B_{\rho\sigma\sigma_0}(x)} \nabla\eta \cdot \nabla d^2(v, \hat{w}) \, d\mu.$$

and since η is supported in $B_{\rho\sigma\sigma_0}(x)$

$$(80) \quad -C \int_{B_{\rho\sigma}(x)} \eta d(v, P_0) d(v, \hat{w}) d\mu \leq -\int_{B_{\rho\sigma}(x)} \nabla\eta \cdot \nabla d^2(v, \hat{w}) \, d\mu.$$

After the change of variables $y = \sigma\zeta$ and multiplication by ν_σ^{-2} we obtain inequality of the assertion. Finally, the monotonicity formula for the energy of the harmonic map w implies that for any $y \in B_{\rho\sigma_0}(0)$,

$$|\nabla w|^2(y) \leq \frac{E_y^w(\rho - \rho\sigma_0)}{(\rho - \rho\sigma_0)^n} \leq \frac{E_0^w(\rho)}{(\rho - \rho\sigma_0)^n} = \frac{E_0^v(\rho)}{(\rho - \rho\sigma_0)^n} \leq \frac{A}{(\rho - \rho\sigma_0)^n}.$$

Thus, the Lipschitz constant of w is only dependent on A and σ_0. Furthermore since

$$|\nabla \hat{w}|(x) = \frac{\nu_\sigma}{\sigma} |\nabla w|(\sigma x)$$
$$= \sqrt{\frac{I^v(\sigma)}{\sigma^{n+1}}} |\nabla w|(\sigma x)$$
$$\leq C |\nabla w|(\sigma x) \quad \text{(by the Lipschitz bound of } u\text{)}$$

we obtain that the constant C in the statement of the lemma is only dependent on on σ_0, the constant in the estimates (28)–(32) for the target metric G, the domain metric g, the Lipschitz constant of u and A as required. □

PROPOSITION 48. *Let* $u = (V, v) : B_{\sigma_\star}(x_\star) \to (\mathbf{R}^j \times Y_2^{k-j}, d_G)$ *be a harmonic map as in (16) and* $x \in \mathcal{S}_j(u) \cap B_{\frac{\sigma_\star}{2}}(x_\star)$. *Identify* $x = 0$ *via normal coordinates and let* $\sigma > 0$ *sufficiently small such that* $B_\sigma(0) \subset B_{\frac{\sigma_\star}{2}}(x_\star)$. *For* v_σ, w_σ *and* g_σ *as in Definition 44, assume there exists* $A > 0$ *such that* $E^{v_\sigma}(1) \leq A$. *Then given* $R \in (0, 1)$ *sufficiently small, there exists* $C > 0$ *depending only on the dimension* n,

7. LOWER ORDER BOUND

the metric g of the domain, the constant in the estimates (28)–(32) for the target metric G, the Lipschitz constant of u, R and A such that

$$\sup_{B_R(0)} d^2(v_\sigma, w_\sigma) \leq C\sigma^2.$$

PROOF. For $x \in B_R(0)$ and $s \in (0, \frac{1-R}{2})$, let w be a map as in Lemma 47 and let η approximate the characteristic function of $B_\rho(x)$ to obtain

$$-C\sigma^2 \int_{B_s(x)} d(v_\sigma, w)d(v_\sigma, P_0) d\mu_\sigma \leq \int_{\partial B_s(x)} \frac{\partial}{\partial s} d^2(v_\sigma, w) \, d\mu_\sigma.$$

Standard computation shows that

$$\int_{\partial B_s(x)} \frac{\partial}{\partial s} d^2(v_\sigma, w) d\Sigma_\sigma$$

$$\leq s^{n-1} \frac{d}{ds} \left(\frac{1}{s^{n-1}} \int_{\partial B_s(x)} d^2(v_\sigma, w) d\Sigma_\sigma \right) + C\sigma^2 s \int_{\partial B_s(x)} d^2(v_\sigma, w) d\Sigma_\sigma$$

$$\leq s^{n-1} \frac{d}{ds} \left(\frac{e^{Cs^2}}{s^{n-1}} \int_{\partial B_s(x)} d^2(v_\sigma, w) d\Sigma_\sigma \right)$$

where the $C\sigma^2$ term comes from the fact that the domain metric is non-Euclidean. Combining the above two inequalities

$$0 \leq \frac{d}{ds} \left(\frac{e^{Cs^2}}{s^{n-1}} \int_{\partial B_s(x)} d^2(v_\sigma, w) d\Sigma_\sigma \right) + C\sigma^2 s^{-n+1} \int_{B_s(x)} d(v_\sigma, w) d(v_\sigma, P_0) d\mu_\sigma$$

$$\leq \frac{d}{ds} \left(\frac{e^{Cs^2}}{s^{n-1}} \int_{\partial B_s(x)} d^2(v_\sigma, w) d\Sigma_\sigma \right)$$

(81) $$+ C\sigma^2 s^{-n+1} \int_{B_s(x)} d^2(v_\sigma, w) d\mu_\sigma + C\sigma^2 s^{-n+1} \int_{B_s(x)} d^2(v_\sigma, P_0) d\mu_\sigma.$$

Now set $w = w_\sigma$. Clearly $E^{w_\sigma}(1) \leq E^{v_\sigma}(1) \leq A$. Furthermore, by Lemma 45 and the assumption $E^{v_\sigma}(1) \leq A$, we thus have that

$$0 \leq \frac{d}{ds} \left(\frac{e^{Cs^2}}{s^{n-1}} \int_{\partial B_s(x)} d^2(v_\sigma, w_\sigma) d\Sigma_\sigma \right) + C\sigma^2 s^{-n+1}.$$

Integrating this over $s \in (0, t)$, we obtain

$$0 \leq \frac{e^{Ct^2}}{t^{n-1}} \int_{\partial B_t(x)} d^2(v_\sigma, w_\sigma) d\Sigma_\sigma - \frac{1}{C_n} d^2(v_\sigma(x), w_\sigma(x)) + C\sigma^2 t^{-n+2}$$

where C_n depends only on n. Thus,

$$t^{n-1} d^2(v_\sigma(x), w_\sigma(x)) \leq C \int_{\partial B_t(x)} d^2(v_\sigma, w_\sigma) d\Sigma_\sigma + C\sigma^2 t.$$

Integrating this over $t \in (0, \frac{1-R}{2})$, we have

$$d^2(v_\sigma(x), w_\sigma(x)) \leq C \int_{B_{\frac{1-R}{2}}(x)} d^2(v_\sigma, w_\sigma) d\mu_\sigma + C\sigma^2$$

$$\leq C \int_{B_1(0)} d^2(v_\sigma, w_\sigma) d\mu_\sigma + C\sigma^2.$$

Thus, the assertion of the lemma follows from Lemma 45 and Lemma 46. □

For u as in Proposition 43, $\sigma_i > 0$ and $x_i \in \mathcal{S}_j(u) \cap B_{\frac{\sigma_\star}{2}}(x_\star)$, use normal coordinates to write the unit ball centered at $x_i = 0$ as $(B_1(0), g)$. (Here, by rescaling if necessary, we can assume without the loss of generality that $B_1(0) \subset B_{\sigma_\star}(x_\star)$.) Define the σ_i-blow up map and the approximating harmonic σ_i-blow up map at x_i as in Definition 44 and denote them as

(82) $\qquad v_i, w_i : (B_1(0), g_i) \to (Y_2^{k-j}, d_h)$ where $g_i(x) = g(\sigma_i x)$.

Furthermore, set
$$\nu_i := \left(\frac{I^v_{x_i}(\sigma_i)}{\sigma_i^{n-1}}\right)^{1/2}.$$

LEMMA 49. *Let $u = (V, v) : B_{\sigma_\star}(x_\star) \to (\mathbf{R}^j \times Y_2^{k-j}, d_G)$ be a harmonic map as in (16), $x_i \in \mathcal{S}_j(u) \cap B_{\frac{\sigma_\star}{2}}(x_\star)$, $\sigma_i \to 0$ and v_i be as in (82). If there exists $A > 0$ such that*

(83) $$\frac{\sigma_i E^v_{x_i}(\sigma_i)}{I^v_{x_i}(\sigma_i)} \leq A$$

then there exists a subsequence of $\{i\}$ (which we denote again by $\{i\}$ by abuse of notation) and a non-constant harmonic map $v_0 : (B_1(0), \delta) \to Y_0$ into an NPC space such that $v_i \to v_0$, $w_i \to v_0$ locally uniformly in the pullback sense. (Here, δ is the Euclidean metric.) Furthermore, (after identifying $x_i = 0$ via normal coordinates)

(84) $\qquad I_0^{v_0}(1) = \lim_{i \to \infty} I_0^{v_i}(1) = 1$ *and* $E_0^{v_0}(1) \leq \lim_{i \to \infty} E_0^{v_i}(1)$.

PROOF. Let w_i as in (82), identify $x_i = 0$ via normal coordinates and write $E = E_0$, $I = I_0$ for simplicity. By Assumption (83) and the energy minimizing property of w_i, we have

(85) $$E^{w_i}(1) \leq E^{v_i}(1) = \frac{E^{v_i}(1)}{I^{v_i}(1)} \leq A.$$

Therefore, w_i is a family of harmonic maps with uniformly bounded energy. For any $r \in (0, 1)$, the Lipschitz constant of w_i in $B_r(0)$ depends only the energy bound and r and is independent of i (cf. [**KS1**] Theorem 2.4.6). Thus, w_i has a locally uniform Lipschitz constant and, by [**KS2**] Proposition 3.7, there exists a subsequence (which we still denote by $\{i\}$ by an abuse of notation) such that w_i converges locally uniformly in the pullback sense to a map v_0. By [**KS2**] Theorem 3.11, v_0 is energy minimizing on $B_r(0)$ for any $r \in (0, 1)$. The fact that v_0 is energy minimizing on every compact subset of $B_1(0)$ immediately implies v_0 is energy minimizing on $B_1(0)$ by the same argument as in (4).

We now claim

(86) $\qquad d(v_i, w_i) \to 0$ in $W^{1,2}$.

To prove (86), first note that by Lemma 46 and (85),

(87) $$E^{v_i}(1) - E^{w_i}(1) \leq C\sigma_i^2.$$

Hence, Lemma 45 implies
$$\int_{B_1(0)} |\nabla d(v_i, w_i)|^2 \, d\mu_i \leq C\sigma_i^2$$

and
$$\tag{88} \int_{B_1(0)} d^2(v_i, w_i) \, d\mu_i \leq C\sigma_i^2.$$

Since $d\mu_i$ is uniformly close to the Euclidean volume form $d\mu_0$ it follows that $d(v_i, w_i) \to 0$ in $W^{1,2}$ as claimed in (86). It now follows from Proposition 48 that $d(v_i, w_i) \to 0$ uniformly in $B_R(0)$, and hence

$$\tag{89} \lim_{i \to \infty} v_i = v_0 \quad \text{uniformly in the pullback sense in } B_R(0).$$

The harmonicity of w_i implies the subharmonicity of $d^2(w_i, P_0)$, and hence
$$\int_{\partial B_r(0)} d^2(w_i, P_0) d\Sigma_i \leq Cr^{n-1} \int_{\partial B_1(0)} d^2(w_i, P_0) d\Sigma_i \leq C.$$

Since $d(P_0, w_i(0)) = d(v_i(0), w_i(0)) \to 0$, we have
$$\lim_{i \to 0} \int_{\partial B_r(0)} d^2(w_i, P_0) d\Sigma_i = \lim_{i \to 0} \int_{\partial B_r(0)} d^2(w_i, w_i(0)) d\Sigma_i = \int_{\partial B_r(0)} d^2(v_0, v_0(0)) d\Sigma_0$$

where $d\Sigma_0$ is the volume form with respect to the Euclidean metric. Thus, by the Dominated Convergence Theorem,
$$\begin{aligned} \lim_{i \to 0} \int_{B_1(0)} d^2(w_i, P_0) d\mu_i &= \int_0^1 \lim_{i \to 0} \int_{\partial B_r(0)} d^2(w_i, P_0) d\Sigma_i dr \\ &= \int_0^1 \int_{\partial B_r(0)} d^2(v_0, v_0(0)) d\Sigma_0 dr \\ &= \int_{B_1(0)} d^2(v_0, v_0(0)) d\mu_0. \end{aligned}$$

Thus, the the L^2 convergence of $d(v_i, w_i)$ to 0 implies,
$$\lim_{i \to 0} \int_{B_1(0)} d^2(v_i, P_0) d\mu_i = \int_{B_1(0)} d^2(v_0, v_0(0)) d\mu_0.$$

Finally, since
$$\int_{B_1(0)} |\nabla d(v_i, P_0)|^2 d\mu_i \leq \int_{B_1(0)} |\nabla v_i|^2 d\mu_i \leq A,$$
we conclude by standard $W^{1,2}$-trace theory that
$$1 = \lim_{i \to \infty} \int_{\partial B_1(0)} d^2(v_i, P_0) d\mu_i = \int_{\partial B_1(0)} d^2(v_0, v_0(0)) d\mu_0$$

which is the first assertion of (84). By uniform Lipschitz continuity of w_i and the lower semicontinuity of energy [**KS2**] Lemma 3.8, we have
$$E_0^{v_0}(1) \leq \lim_{i \to \infty} E_0^{w_i}(1).$$

Combined with (87), we obtain the second assertion of (84). □

PROOF OF PROPOSITION 43. If (72) is not true, then there exist sequences $x_i \in \mathcal{S}_j(u) \cap B_{\frac{\sigma_*}{2}}(x_*)$ and $\sigma_i \to 0$ such that
$$\frac{\sigma_i E_{x_i}^v(\sigma_i)}{I_{x_i}^v(\sigma_i)} < 1 - \epsilon_0$$

which is equivalent to
$$\frac{E^{v_i}(1)}{I^{v_i}(1)} < 1 - \epsilon_0.$$
By (84),
$$\frac{E^{v_0}(1)}{I^{v_0}(1)} \leq \lim_{i \to 0} \frac{E^{v_i}(1)}{I^{v_i}(1)} \leq 1 - \epsilon_0.$$
On the other hand, since v_0 is a nonconstant harmonic map with respect to the Euclidean metric, it follows that
$$1 \leq \frac{E^{v_0}(1)}{I^{v_0}(1)}.$$
The contradiction proves the assertion of the Proposition. □

CHAPTER 8

The Domain variation

The main goal of this section is to obtain estimates for the domain variation of the singular component map $v : B_{\sigma_*}(x_*) \to (Y_2^{k-j}, d_h)$. We start by showing a regularity result for the non-singular component map.

LEMMA 50. *Let* $u = (V, v) : B_{\sigma_*}(x_*) \to (\mathbf{R}^j \times Y_2^{k-j}, d_G)$ *be a harmonic map as in (16). If* $x_0 \in B_{\frac{\sigma_*}{2}}(x_*)$ *and* $\sigma \in (0, \frac{\sigma_*}{2})$, *then* $V^I \in W^{2,p}(B_\sigma(x_0))$ *for any* $p > 1$.

PROOF. For a smooth $\eta = (\eta^1, \ldots, \eta^j)$ with compact support in $B_\sigma(x_0)$, let $V_t = V + t\eta$ and $u_t = (V_t, v)$. Assumption 5 states $|\nabla v|^2(x) = 0$ for almost every $x \in \mathcal{S}_j(u)$, and hence

$$\begin{aligned}|\nabla u_t|^2(x) &= |\nabla V_t|^2(x) \\ &\quad - \mathbf{G}_{11}(V_t, v)\nabla V_t \cdot \nabla V_t(x) \\ &= \mathbf{G}_{11}(V_t, v)\nabla V \cdot \nabla V(x) + 2t\mathbf{G}_{11}(V_t, v)\nabla V \cdot \nabla \eta(x) \\ &\quad + t^2 \mathbf{G}_{11}(V_t, v)\nabla \eta \cdot \nabla \eta(x).\end{aligned}$$

In $\mathcal{R}(u)$,

$$\begin{aligned}|\nabla u_t|^2 &= \mathbf{G}_{11}(V_t, v)\nabla V_t \cdot \nabla V_t + 2\mathbf{G}_{12}(V_t, v)\nabla V_t \cdot \nabla v + \mathbf{G}_{22}(V_t, v)\nabla v \cdot \nabla v \\ &= \mathbf{G}_{11}(V_t, v)\nabla V \cdot \nabla V + 2t\mathbf{G}_{11}(V_t, v)\nabla V \cdot \nabla \eta + t^2 \mathbf{G}_{11}(V_t, v)\nabla \eta \cdot \nabla \eta \\ &\quad + 2\mathbf{G}_{12}(V_t, v)\nabla V \cdot \nabla v + 2t\mathbf{G}_{12}(V_t, v)\nabla \eta \cdot \nabla v + \mathbf{G}_{22}(V_t, v)\nabla v \cdot \nabla v.\end{aligned}$$

Thus, $|\nabla u_t|^2(x)$ is an integrable function in the variables x, t and, for almost every $x \in B_\sigma(x_0)$, $|\nabla u_t|^2(x)$ is a smooth function in t. Furthermore, $\frac{d}{dt}|\nabla u_t|^2$ is bounded independently of t by an L^1 function by the metric estimates and the Lipschitz continuity of u. We can thus conclude that $t \mapsto E(u_t)$ is a smooth function in t, and its derivatives can be computed by differentiation under the integral sign. In particular, since $\frac{d}{dt}E(u_t)\big|_{t=0} = 0$, we obtain

$$
\begin{aligned}
0 &= \int_{B_\sigma(x_0)} \frac{d}{dt}\bigg|_{t=0} \mathbf{G}_{11}(V_t,v) \nabla V \cdot \nabla V + 2\mathbf{G}_{11}(V,v) \nabla V \cdot \nabla \eta \, d\mu \\
&\quad + 2 \int_{B_\sigma(x_0) \cap \mathcal{R}(u)} \frac{d}{dt}\bigg|_{t=0} \mathbf{G}_{12}(V_t,v)\big|_{t=0} \nabla V \cdot \nabla v + \mathbf{G}_{12}(V,v) \nabla \eta \cdot \nabla v \, d\mu \\
&\quad + 2 \int_{B_\sigma(x_0) \cap \mathcal{R}(u)} \frac{d}{dt}\bigg|_{t=0} \mathbf{G}_{22}(V_t,v)\big|_{t=0} \nabla v \cdot \nabla v \, d\mu \\
&= \int_{B_\sigma(x_0)} \eta^I \frac{\partial}{\partial V^I} \mathbf{G}_{11}(V,v) \nabla V \cdot \nabla V + 2\mathbf{G}_{11}(V,v) \nabla V \cdot \nabla \eta \, d\mu \\
&\quad + 2 \int_{B_\sigma(x_0) \cap \mathcal{R}(u)} \eta^I \frac{\partial}{\partial V^I} \mathbf{G}_{12}(V,v) \nabla V \cdot \nabla v + \mathbf{G}_{12}(V,v) \nabla \eta \cdot \nabla v \, d\mu \\
&\quad + \int_{B_\sigma(x_0) \cap \mathcal{R}(u)} \eta^I \frac{\partial}{\partial V^I} \mathbf{G}_{22}(V,v) \nabla v \cdot \nabla v \, d\mu.
\end{aligned}
\tag{90}
$$

By applying integration by parts in the same way as the term $(II)_2$ of Proposition 37, we obtain

$$
\begin{aligned}
\int_{B_\sigma(x_0) \cap \mathcal{R}(u)} \mathbf{G}_{12}(V,v) \nabla \eta \cdot \nabla v \, d\mu &= \int_{B_\sigma(x_0) \cap \mathcal{R}(u)} g^{\alpha\beta} G_{Ik}(V,v) \frac{\partial \eta^I}{\partial x^\alpha} \frac{\partial v^k}{\partial x^\beta} d\mu \\
&= \int_{B_\sigma(x_0)} \eta^I f_{Ik} d\mu
\end{aligned}
\tag{91}
$$

where f_{Ik} is a bounded function. Thus, (90) implies that

$$
-\int_{B_\sigma(x_0)} g^{\alpha\beta} G_{IJ}(V,v) \frac{\partial V^I}{\partial x^\alpha} \frac{\partial \eta^J}{\partial x^\beta} d\mu = \int_{B_\sigma(x_0)} \eta \cdot F d\mu
\tag{92}
$$

for some bounded vector field F. Let

$$
\eta^J = \sum_K G^{JK}(V,v) \varphi
$$

for $\varphi \in C_c^\infty(B_\sigma(x_0))$. Then

$$
\frac{\partial \eta^J}{\partial x^\beta} = \sum_K \left(\varphi \frac{\partial}{\partial V^L} G^{JK}(V,v) \frac{\partial V^L}{\partial x^\beta} + \varphi \frac{\partial}{\partial v^l} G^{JK}(V,v) \frac{\partial v^l}{\partial x^\beta} \right. \\
\left. + G^{JK}(V,v) \frac{\partial \varphi}{\partial x^\beta} \right)
$$

and hence

$$
G_{IJ}(V,v) \frac{\partial \eta^J}{\partial x^\beta} = \varphi \sum_K \left(G_{IJ}(V,v) \frac{\partial}{\partial V^L} G^{JK}(V,v) \frac{\partial V^L}{\partial x^\beta} \right. \\
\left. + G_{IJ}(V,v) \frac{\partial}{\partial v^l} G^{JK}(V,v) \frac{\partial v^l}{\partial x^\beta} \right) + \frac{\partial \varphi}{\partial x^\beta}.
$$

Since H is a smooth Riemannian metric, $H_{II}^{\frac{1}{2}}, H_{KK}^{-\frac{1}{2}}$ are uniformly bounded. Thus, (28), (29), (30) and (47) imply

$$\left| G_{IJ} \frac{\partial}{\partial V^L} G^{JK} \frac{\partial V^L}{\partial x^\beta} \right| = \left| G_{IJ} G^{JM} \frac{\partial}{\partial V^L} G_{MN} G^{NK} \frac{\partial V^L}{\partial x^\beta} \right|$$
$$\leq H_{II}^{\frac{1}{2}} H_{KK}^{-\frac{1}{2}} \left| H_{LL}^{\frac{1}{2}} \frac{\partial V^L}{\partial x^\beta} \right|$$
$$\leq C$$

and

$$\left| G_{IJ}(V,v) \frac{\partial}{\partial v^l} G^{JK}(V,v) \frac{\partial v^l}{\partial x^\beta} \right| = \left| G_{IJ} G^{JM} \frac{\partial}{\partial V^L} G_{MN} G^{NJ} \right|$$
$$\leq H_{II}^{\frac{1}{2}} H_{KK}^{-\frac{1}{2}} \left| h_{ll}^{\frac{1}{2}} \frac{\partial v^l}{\partial x^\beta} \right|$$
$$\leq C.$$

Thus, (92) implies

$$-\int_{B_\sigma(x_0)} g^{\alpha\beta} \frac{\partial V^I}{\partial x^\alpha} \frac{\partial \varphi}{\partial x^\beta} d\mu = \int_{B_\sigma(x_0)} \varphi \cdot f d\mu$$

for some bounded function f. By elliptic regularity, $V^I \in W^{2,p}(B_\sigma(x_0))$. □

We now prove the following weaker version of the Lemma 50 for u and the singular component map v.

LEMMA 51. *Let $u = (V,v) : B_{\sigma_*}(x_*) \to (\mathbf{R}^j \times Y_2^{k-j}, d_G)$ be a harmonic map as in (16). If $x_0 \in B_{\frac{\sigma_*}{2}}(x_*)$ and $\sigma \in (0, \frac{\sigma_*}{4})$, then there exists a constant $C > 0$ depending only on the dimension of the domain, the metric g and the total energy of u such that*

$$\int_{B_\sigma(x_0) \setminus \{d(v,P_0)=0\}} d(v,P_0) |\nabla \nabla u| d\mu \leq C$$

and

$$\int_{B_\sigma(x_0) \setminus \{d(v,P_0)=0\}} d(v,P_0) |\nabla \nabla v| d\mu \leq C.$$

PROOF. Let

$$d_\epsilon = \max\{d(v,P_0) - \epsilon, 0\}$$

and $\varphi \in C_c^\infty(B_{\frac{\sigma_*}{2}}(x_0))$ such that $0 \leq \varphi \leq 1$, $\varphi = 1$ on $B_\sigma(x_0)$, $\varphi = 0$ outside $B_{\frac{3\sigma_*}{8}}(x_0)$ and $|\nabla \varphi| \leq \frac{16}{\sigma_*}$. Let Ω_1 be the support of the function $d_\epsilon^2 \varphi^2$ which is compactly contained in $B_{\frac{\sigma_*}{2}}(x_0) \setminus \{d(v,P_0) = 0\} \subset B_{\frac{\sigma_*}{2}}(x_0) \setminus \mathcal{S}_j(u)$. By the proof of [GS] Lemma 6.6, Assumption 6 implies that the inequality

$$\frac{1}{2} \triangle |\nabla u|^2 \geq |\nabla \nabla u|^2 - c|\nabla u|^2$$

holds distributionally in Ω_1. Thus by using $d_\epsilon^2 \varphi^2$ as a the test function

$$-\int_{B_{\frac{\sigma_*}{2}}(x_0)} d_\epsilon \varphi \nabla(d_\epsilon \varphi) \cdot \nabla |\nabla u|^2 d\mu \geq \int_{B_{\frac{\sigma_*}{2}}(x_0)} d_\epsilon^2 \varphi^2 (|\nabla \nabla u|^2 - c|\nabla u|^2) d\mu.$$

After an application of the arithmetic-geometric means inequality, we obtain

$$\frac{1}{2}\int_{B_{\frac{\sigma_\star}{2}}(x_0)} |\nabla(d_\epsilon \varphi)|^2 |\nabla u|^2 d\mu + c \int_{B_{\frac{\sigma_\star}{2}}(x_0)} d_\epsilon^2 \varphi^2 |\nabla u|^2 d\mu \geq \frac{1}{2}\int_{B_{\frac{\sigma_\star}{2}}(x_0)} d_\epsilon^2 \varphi^2 |\nabla\nabla u|^2 d\mu.$$

Noting that $d_\epsilon^2 \varphi^2$, $\varphi^2|\nabla d_\epsilon|^2$ are bounded by the Lipschitz constant of v (and hence of u) in $B_{\frac{\sigma_\star}{2}}(x_0)$, we obtain,

$$\left(\int_{B_\sigma(x_0)} d_\epsilon |\nabla\nabla u| d\mu\right)^2$$

$$\leq C \int_{B_\sigma(x_0)} d_\epsilon^2 |\nabla\nabla u|^2 d\mu$$

$$\leq C \int_{B_{\frac{\sigma_\star}{2}}(x_0)} d_\epsilon^2 \varphi^2 |\nabla\nabla u|^2 d\mu$$

$$\leq C \left(\int_{B_{\frac{\sigma_\star}{2}}(x_0)} |\nabla(d_\epsilon\varphi)|^2 |\nabla u|^2 d\mu + 2c \int_{B_{\frac{\sigma_\star}{2}}(x_0)} d_\epsilon^2 \varphi^2 |\nabla u|^2 d\mu \right)$$

$$\leq C.$$

By letting $\epsilon \to 0$, the first inequality follows. The second inequality follows from the first. □

Let $u = (V,v) : (B_{\sigma_\star}(x_\star), g) \to (\mathbf{R}^j \times Y_2^{k-j}, d_G)$ be a harmonic map satisfying the assumptions of Section 5, $x_0 \in \mathcal{S}_j(u) \cap B_{\frac{\sigma_\star}{2}}(x_\star)$ and let $r_0 \in (0, \frac{\sigma_\star}{4})$. Define the map $v_t : B_{r_0}(x_0) \to (Y_2^{k-j}, d_h)$ by setting

$$v_t(x) = v \circ F_t(x)$$

where F_t is a diffeomorphism given by

$$F_t(x) = (1 + t\xi(x))x, \quad \xi \in C_c^\infty(B_{r_0}(x_0)), \quad 0 \leq \xi \leq 1.$$

Define

$$u_t : B_{r_0}(x_0) \to (\mathbf{R}^j \times Y_2^{k-j}, d_G)$$

by setting

$$u_t := (V, v_t).$$

Since $u = u_t$ on $\partial B_\sigma(x_0)$, u_t is a competitor.

LEMMA 52. *Let $u = (V,v) : B_{\sigma_\star}(x_\star) \to (\mathbf{R}^j \times Y_2^{k-j}, d_G)$ be a harmonic map as in (16). There exists $C > 0$ such that for $x_0 \in \mathcal{S}_j(u) \cap B_{\frac{\sigma_\star}{2}}(x_\star)$ and $\sigma \in (0, r_0)$, we have*

$$\lim_{t \to 0} \frac{E_{x_0}^v(\sigma) - E_{x_0}^{v_t}(\sigma)}{t} \leq C \int_{B_\sigma(x_0)} \xi d^2(v, P_0) d\mu + C\sigma \int_{B_\sigma(x_0)} \xi |\nabla v|^2 d\mu$$

Furthermore, C depends only on the constant in the estimates (28)–(32) for the target metric G, the domain metric g and the Lipschitz constant of u.

PROOF. First note that since $v \in W^{1,2}$, the same argument as in [**GS**] p.192 implies that the limit on the left hand side of the inequality above exists. Moreover, we can take the limit under the integral sign to obtain

(93) $$\lim_{t\to 0} \frac{E^v_{x_0}(\sigma) - E^{v_t}_{x_0}(\sigma)}{t}$$

$$= \int_{B_\sigma(x_0)} \lim_{t\to 0} \frac{|\nabla v|^2 - |\nabla v_t|^2}{t} d\mu$$

$$= \int_{\mathcal{R}(u)\cap B_\sigma(x_0)} \lim_{t\to 0} \frac{|\nabla v|^2 - |\nabla v_t|^2}{t} d\mu$$

$$+ \int_{\mathcal{S}(u)\cap B_\sigma(x_0)} \lim_{t\to 0} \frac{|\nabla v|^2 - |\nabla v_t|^2}{t} d\mu.$$

Next, we claim

(94) $$\lim_{t\to 0} \frac{1}{t}\int_{B_\sigma(x_0)} |\nabla u_t|^2 - |\nabla u|^2 d\mu = \int_{B_\sigma(x_0)} \lim_{t\to 0} \frac{|\nabla u_t|^2 - |\nabla u|^2}{t} d\mu.$$

We now prove this claim. For almost every $x \in F_t^{-1}(\mathcal{S}_j(u))$, by the chain rule (cf. [KS1] (2.3iv)) and Assumption 5, we have

(95) $$|\nabla v_t|^2(x) = 0 \quad \text{and} \quad |\nabla u_t|^2(x) = |\nabla V|^2(x).$$

By Assumption 3 (ii), this implies that for almost every $x \in F_t^{-1}(\mathcal{S}_j(u))$, we can write by letting $y = F_t(x)$

(96) $$|\nabla u_t|^2(x) = |\nabla V|^2(x)$$

$$= \mathbf{G}_{11}(V(x), v_t(x))\nabla V \cdot \nabla V(x)$$

$$= g^{\alpha\beta}(F_t^{-1}(y))\mathbf{G}_{11}(V(F_t^{-1}(y)),$$

$$v(F_t^{-1}(y)))_{IJ}\frac{\partial V^I}{\partial x^\alpha}(F_t^{-1}(y))\frac{\partial V^J}{\partial x^\beta}(F_t^{-1}(y)).$$

For $x \in F_t^{-1}(\mathcal{R}(u))$, again let $y = F_t(x)$ and write

(97) $$|\nabla u_t|^2(x) = \mathbf{G}_{11}(V(x), v_t(x))\nabla V \cdot \nabla V(x)$$

$$+ 2\mathbf{G}_{12}(V(x), v_t(x))\nabla V \cdot \nabla v_t(x) + \mathbf{G}_{22}(V(x), v_t(x))\nabla v_t \cdot \nabla v_t$$

$$= g^{\alpha\beta}(F_t^{-1}(y))\mathbf{G}_{11}(V(F_t^{-1}(y)), v(y))_{IJ}$$

$$\times \frac{\partial V^I}{\partial x^\alpha}(F_t^{-1}(y))\frac{\partial V^J}{\partial x^\beta}(F_t^{-1}(y))$$

$$+ 2g^{\alpha\beta}(F_t^{-1}(y))\mathbf{G}_{12}(V(F_t^{-1}(y)), v(y))_{Il}\frac{\partial V^I}{\partial x^\alpha}(F_t^{-1}(y))$$

$$\cdot \frac{\partial v^l}{\partial y^\gamma}(y)\frac{\partial y^\gamma}{\partial x^\beta}(F_t^{-1}(y))$$

$$+ g^{\alpha\beta}(F_t^{-1}(y))\mathbf{G}_{22}(V(F_t^{-1}(y)), v(y))_{lm}\frac{\partial v^l}{\partial y^\gamma}(y)\frac{\partial y^\gamma}{\partial x^\alpha}(F_t^{-1}(y))$$

$$\times \frac{\partial v^m}{\partial y^\delta}(y)\frac{\partial y^\delta}{\partial x^\beta}(F_t^{-1}(y)).$$

Thus, $|\nabla u_t|^2(x)$ is an integrable function in the variables x, t and, for almost every $x \in B_\sigma(x_0)$, $|\nabla u_t|^2(x)$ is a smooth function in t. Furthermore, $\frac{d}{dt}|\nabla u_t|^2$ involves only second derivatives of V and first derivatives of v. Hence, $\frac{d}{dt}|\nabla u_t|^2$ is bounded independently of t by an L^1 function by the metric estimates (28),

(29), the Lipschitz continuity of u and Lemma 50. We can thus conclude that the derivative of $t \mapsto E(u_t)$ can be computed by differentiation under the integral sign. This proves (94).

Since u is harmonic,

$$0 = \lim_{t \to 0} \frac{E^u_{x_0}(\sigma) - E^{u_t}_{x_0}(\sigma)}{t}$$

$$(98) = \lim_{t \to 0} \frac{1}{t} \int_{B_\sigma(x_0)} |\nabla u|^2 - |\nabla u_t|^2 \, d\mu$$

$$= \int_{B_\sigma(x_0)} \lim_{t \to 0} \frac{|\nabla u|^2 - |\nabla u_t|^2}{t} \, d\mu \quad \text{by (94)}$$

$$= \int_{\mathcal{R}(u) \cap B_\sigma(x_0)} \lim_{t \to 0} \frac{|\nabla u|^2 - |\nabla u_t|^2}{t} \, d\mu + \int_{\mathcal{S}(u) \cap B_\sigma(x_0)} \lim_{t \to 0} \frac{|\nabla u|^2 - |\nabla u_t|^2}{t} \, d\mu.$$

To address the integral over $\mathcal{S}_j(u) \cap B_\sigma(x_0)$ on the right hand side above, we consider the following two sets $\mathcal{S}_j(u) \cap F_t^{-1}(\mathcal{S}_j(u))$ and $\mathcal{S}_j(u) \cap F_t^{-1}(\mathcal{R}_j(u))$. By Lemma 29 and (95),

$$\frac{|\nabla u|^2(x) - |\nabla u_t|^2(x)}{t} = 0 = \frac{|\nabla v|^2(x) - |\nabla v_t|^2(x)}{t}$$

for almost every $x \in \mathcal{S}_j(u) \cap F_t^{-1}(\mathcal{S}_j(u))$. For $x \in \mathcal{S}_j(u) \cap F_t^{-1}(\mathcal{R}(u))$,

$$d(v_t(x), P_0) = d(v_t(x), v(x)) \leq C|F_t(x) - x| \leq Ct\xi(x)|x|,$$

and hence the metric estimates (28) imply

$$|\nabla u_t|^2(x)$$
$$= \mathbf{G}_{11}(V, v_t)\nabla V \cdot \nabla V + 2\mathbf{G}_{12}(V, v_t)\nabla V \cdot \nabla v_t + \mathbf{G}_{22}(V, v_t)\nabla v_t \cdot \nabla v_t$$
$$= |\nabla V|^2(x) + |\nabla v_t|^2(x) + O(t^2).$$

Thus, for almost every $x \in \mathcal{S}_j(u)$, we have

$$(99) \quad \left| \frac{|\nabla u_t|^2(x) - |\nabla u|^2(x)}{t} - \frac{|\nabla v_t|^2(x) - |\nabla v|^2(x)}{t} \right| \leq O(t).$$

Since $\mathcal{S}_j(u)$ is of full measure in $\mathcal{S}(u)$ by Assumption 3 (ii), (98) and (99) imply

$$\int_{\mathcal{S}(u) \cap B_\sigma(x_0)} \lim_{t \to 0} \frac{|\nabla v|^2 - |\nabla v_t|^2}{t} \, d\mu = \int_{\mathcal{R}(u) \cap B_\sigma(x_0)} \lim_{t \to 0} \frac{|\nabla u_t|^2 - |\nabla u|^2}{t} \, d\mu.$$

Combined with (93), we obtain

$$(100) \lim_{t \to 0} \frac{E^v_{x_0}(\sigma) - E^{v_t}_{x_0}(\sigma)}{t}$$

$$= -\int_{\mathcal{R}(u) \cap B_\sigma(x_0)} \lim_{t \to 0} \frac{d}{dt}\Big|_{t=0} |\nabla v_t|^2 d\mu + \int_{\mathcal{R}(u) \cap B_\sigma(x_0)} \frac{d}{dt}\Big|_{t=0} |\nabla u_t|^2 d\mu.$$

For $x \in \mathcal{R}(u)$ and t sufficiently small such that $F_t(x) \in \mathcal{R}(u)$,

$$|\nabla u_t|^2 - |\nabla u|^2 = \mathbf{G}_{11}(V, v_t)\nabla V \cdot \nabla V - \mathbf{G}_{11}(V, v)\nabla V \cdot \nabla V$$
$$+ 2(\mathbf{G}_{12}(V, v_t)\nabla V \cdot \nabla v_t - \mathbf{G}_{12}(V, v)\nabla V \cdot \nabla v)$$
$$+ \mathbf{G}_{22}(V, v_t)\nabla v_t \cdot \nabla v_t - \mathbf{G}_{22}(V, v)\nabla v \cdot \nabla v.$$

8. THE DOMAIN VARIATION

Divide the above by t and take the limit as $t \to 0$. Integrating the resulting inequality and combining with (100)

$$\lim_{t \to 0} \frac{E^v_{x_0}(\sigma) - E^{v_t}_{x_0}(\sigma)}{t} = \int_{\mathcal{R}(u) \cap B_\sigma(x_0)} \frac{d}{dt}\bigg|_{t=0} \mathbf{G}_{11}(V, v_t) \nabla V \cdot \nabla V \, d\mu$$
$$+ 2 \int_{\mathcal{R}(u) \cap B_\sigma(x_0)} \frac{d}{dt}\bigg|_{t=0} \mathbf{G}_{12}(V, v_t) \nabla V \cdot \nabla v_t \, d\mu$$
$$+ \int_{\mathcal{R}(u) \cap B_\sigma(x_0)} \frac{d}{dt}\bigg|_{t=0} \Box(V, v_t) \nabla v_t \cdot \nabla v_t \, d\mu$$
(101) $\qquad =: (i) + (ii) + (iii)$

where $\Box(V, v) = \mathbf{G}_{22}(V, v) - h(v)$. We claim that

$$(i) := \int_{\mathcal{R}(u) \cap B_\sigma(x_0)} \frac{d}{dt}\bigg|_{t=0} \mathbf{G}_{11}(V, v_t) \nabla V \cdot \nabla V$$
(102) $\qquad \leq C \int_{B_\sigma(x_0)} \xi d^2(v, P_0) d\mu + C\sigma^2 \int_{B_\sigma(x_0)} \xi |\nabla v|^2 d\mu,$

$$(ii) := 2 \int_{\mathcal{R}(u) \cap B_\sigma(x_0)} \frac{d}{dt}\bigg|_{t=0} \mathbf{G}_{12}(V, v_t) \nabla V \cdot \nabla v_t \, d\mu$$
(103) $\qquad \leq C \int_{B_\sigma(x_0)} \xi d^2(v, P_0) d\mu + C\sigma^2 \int_{B_\sigma(x_0)} \xi |\nabla v|^2.$

and

$$(iii) := \int_{\mathcal{R}(u) \cap B_\sigma(x_0)} \frac{d}{dt}\bigg|_{t=0} \Box(V, v_t) \nabla v_t \cdot \nabla v_t \, d\mu$$
(104) $\qquad \leq C \int_{B_\sigma(x_0)} \xi d^2(v, P_0) d\mu + C\sigma \int_{B_\sigma(x_0)} \xi |\nabla v|^2 d\mu.$

Combined with (101), the estimates (102), (103) and (104) prove the Lemma. Thus, our goal now is to prove these estimates.

We first prove (i). Let $x \in \mathcal{R}(u) \cap B_\sigma(x_0)$. Then with

$$y^\alpha = (1 + t\xi(x))x^\alpha \text{ and } \frac{\partial y^\alpha}{\partial t} = \xi(x) x^\alpha,$$

we have

$$\left| \frac{d}{dt}\bigg|_{t=0} \mathbf{G}_{11}(V, v_t)_{IJ} \nabla V^I \cdot \nabla V^J \right| \leq C \left| \sum_{i=1}^{k-j} \frac{\partial}{\partial v^i} \mathbf{G}_{11}(V, v) \sum_\alpha \frac{\partial v^i}{\partial y^\alpha}(x) \frac{\partial y^\alpha}{\partial t} \right|$$
$$\leq C\sigma \xi d(v, P_0) |\nabla v|$$

which in turn implies

$$(i) := \int_{\mathcal{R}(u) \cap B_\sigma(x_0)} \frac{d}{dt}\bigg|_{t=0} \mathbf{G}_{11}(V, v_t) \nabla V \cdot \nabla V$$
$$\leq C\sigma \int_{B_\sigma(x_0)} \xi d(v, P_0) |\nabla v|$$
$$\leq C \int_{B_\sigma(x_0)} \xi d^2(v, P_0) d\mu + C\sigma^2 \int_{B_\sigma(x_0)} \xi |\nabla v|^2 d\mu.$$

This proves (102).

Next, we prove (ii). First, we write

$$
\begin{aligned}
(ii) &:= 2\int_{\mathcal{R}(u)\cap B_\sigma(x_0)} \frac{d}{dt}\bigg|_{t=0} \mathbf{G}_{12}(V,v_t)\nabla V \cdot \nabla v_t d\mu \\
&= 2\int_{\mathcal{R}(u)\cap B_\sigma(x_0)} \frac{d}{dt}\bigg|_{t=0} \mathbf{G}_{12}(V,v_t)\nabla V \cdot \nabla v d\mu \\
&\quad + 2\int_{\mathcal{R}(u)\cap B_\sigma(x_0)} \mathbf{G}_{12}(V,v) \frac{d}{dt}\bigg|_{t=0} \nabla V \cdot \nabla v_t d\mu \\
&=: (ii)_1 + (ii)_2.
\end{aligned}
\tag{105}
$$

We can estimate $(ii)_1$ in similar way as (i) to obtain

$$
\begin{aligned}
(ii)_1 &:= 2\int_{\mathcal{R}(u)\cap B_\sigma(x_0)} \frac{d}{dt}\bigg|_{t=0} \mathbf{G}_{12}(V,v_t)\nabla V \cdot \nabla v d\mu \\
&\leq C\int_{B_\sigma(x_0)} \xi d^2(v,P_0) d\mu + C\sigma^2 \int_{B_\sigma(x_0)} \xi|\nabla v|^2 d\mu.
\end{aligned}
\tag{106}
$$

(Note that in comparison to (i) we used the Lipschitz property of v in order to throw out one term of $|\nabla v|$). We now estimate $(ii)_2$. First, note that since

$$\frac{\partial v_t^i}{\partial x^\beta}(x) = \frac{\partial v^i}{\partial y^\gamma}(y)\left((1+t\xi(x))\delta_{\beta\gamma} + tx^\gamma \frac{\partial \xi}{\partial x^\beta}(x)\right),$$

we also have

$$\frac{d}{dt}\frac{\partial v_t^i}{\partial x^\beta}\bigg|_{t=0} = \frac{\partial^2 v^i}{\partial x^\beta \partial x^\delta}\xi x^\delta + \frac{\partial v^i}{\partial x^\beta}\xi + \frac{\partial v^i}{\partial x^\gamma}x^\gamma \frac{\partial \xi}{\partial x^\beta}$$

hence Lemma 51 implies

$$\int_{B_\sigma(x_0)\setminus\{d(v,P_0)=0\}} d(v,P_0)\left|g^{\alpha\beta}\frac{d}{dt}\frac{\partial v_t^i}{\partial x^\beta}\bigg|_{t=0} h_{ii}^{\frac{1}{2}}\right| d\mu \leq C.$$

Thus, by the metric estimates (28), the Lipschitz property of V^I and with $A_{\epsilon_i}^+$ defined as in the proof of Proposition 37, we have

$$
\begin{aligned}
\int_{B_\sigma(x_0)\setminus(A_{\epsilon_i}^+\cup \mathcal{S}_j(u))} &\left|g^{\alpha\beta}G_{Ii}(V,v)\frac{\partial V^I}{\partial x^\alpha}\frac{d}{dt}\frac{\partial v_t^i}{\partial x^\beta}\bigg|_{t=0}\right| d\mu \\
&\leq C\int_{B_\sigma(x_0)\setminus(A_{\epsilon_i}^+\cup \mathcal{S}_j(u))} d^2(v,P_0)\left|\frac{d}{dt}\frac{\partial v_t^i}{\partial x^\beta}\bigg|_{t=0}\right| d\mu \\
&\leq C\epsilon_i \int_{B_\sigma(x_0)\setminus\{d(v,P_0)=0\}} d(v,P_0)\left|\frac{d}{dt}\frac{\partial v_t^i}{\partial x^\beta}\bigg|_{t=0}\right| d\mu \\
&\leq C\epsilon_i.
\end{aligned}
$$

Thus,

$$(ii)_2 \leq \int_{A_{\epsilon_i}^+} g^{\alpha\beta}G_{Ii}(V,v)\frac{\partial V^I}{\partial x^\alpha}\frac{d}{dt}\frac{\partial v_t^i}{\partial x^\beta}\bigg|_{t=0} d\mu + C\epsilon_i.$$

8. THE DOMAIN VARIATION

Integrating by parts as in (57), we write

$$\int_{A_{e_i}^+} g^{\alpha\beta} G_{Ii}(V,v) \frac{\partial V^I}{\partial x^\alpha} \frac{d}{dt} \frac{\partial v_t^i}{\partial x^\beta}\bigg|_{t=0} d\mu$$

$$= \lim_{\varrho \to 0}\left[-\int_{A_{e_j}^+} \varphi_\varrho \frac{1}{\sqrt{g}} \frac{\partial}{\partial x^\beta}(\sqrt{g} g^{\alpha\beta} \frac{\partial V^I}{\partial x^\alpha}) G_{Ii}(V,v) \frac{dv_t^i}{dt}\bigg|_{t=0} d\mu\right.$$

$$-\int_{A_{e_j}^+} \varphi_\varrho g^{\alpha\beta} \frac{\partial}{\partial x^\beta} G_{Ii}(V,v) \frac{\partial V^I}{\partial x^\alpha} \frac{dv_t^i}{dt}\bigg|_{t=0} d\mu$$

$$-\int_{A_{e_j}^+} g^{\alpha\beta} G_{Ii}(V,v) \frac{\partial V^I}{\partial x^\alpha} \frac{\partial \varphi_\varrho}{\partial x^\beta} \frac{dv_t^i}{dt}\bigg|_{t=0} d\mu$$

$$+\left.\int_{\partial A_{e_j}^+} \varphi_\varrho g^{\alpha\beta} G_{Ii}(V,v) \frac{\partial V^I}{\partial x^\alpha} \frac{dv_t^i}{dt}\bigg|_{t=0} \left(\vec{n} \cdot \frac{\partial}{\partial x^\beta}\right) d\Sigma\right]$$

(107) $$=: \lim_{\varrho \to 0}[(ii)_{21} + (ii)_{22} + (ii)_{23} + (ii)_{24}].$$

By following the proof of estimate $(II)_2$, we obtain

$$(ii)_2 \leq C\sigma \int_{B_\sigma(x_0)} \xi d(v, P_0)|\nabla v| d\mu$$

(108) $$\leq C \int_{B_\sigma(x_0)} \xi d^2(v, P_0) d\mu + C\sigma^2 \int_{B_\sigma(x_0)} \xi |\nabla v|^2 d\mu.$$

Note that we have $\frac{dv_t^i}{dt}\big|_{t=0} = \xi \frac{\partial v^i}{\partial x^\epsilon} x^\epsilon$ in (107) instead of $\frac{dv_{t\eta}^j}{dt}\big|_{t=0} = \eta d(v,w)$ in the corresponding expression (57) for $(II)_2$. This accounts for the difference of $d(v,w)$ and $|\nabla v|$ in the two estimates. We obtain (103) by combining (105), (106) and (108) and Cauchy-Schwartz.

Finally, we estimate (iii). We have

$$(iii) := \int_{\mathcal{R}(u) \cap B_\sigma(x_0)} \frac{d}{dt}\bigg|_{t=0} \Box(V, v_t) \nabla v_t \cdot \nabla v_t \, d\mu$$

$$= \int_{\mathcal{R}(u) \cap B_\sigma(x_0)} \frac{d}{dt}\bigg|_{t=0} \Box(V, v_t) \nabla v \cdot \nabla v \, d\mu$$

(109) $$\quad + 2\int_{\mathcal{R}(u) \cap B_\sigma(x_0)} \Box(V, v) \frac{d}{dt}\bigg|_{t=0} \nabla v_t \cdot \nabla v \, d\mu$$

$$= (iii)_1 + (iii)_2.$$

We derive an estimate for $(iii)_1$ in a similar way as in $(III)_1$ to account for the difference in the C^1 estimates for $\Box(V,v)$ from that of $\mathbf{G}_{12}(V,v)$. We obtain

$$(iii)_1 := \int_{B_\sigma(x_0) \setminus \mathcal{S}_j(u)} g^{\alpha\beta} \frac{\partial}{\partial v^l} \Box_{ij}(V,v) \frac{dv_t^l}{dt}\bigg|_{t=0} \frac{\partial v^i}{\partial x^\alpha} \frac{\partial v^j}{\partial x^\beta} d\mu$$

$$\leq C \int_{B_\sigma(x_0) \setminus \mathcal{S}_j(u)} \left|\xi h_{il}^{\frac{1}{2}} \frac{\partial v^l}{\partial x^\epsilon} x^\epsilon\right| |\nabla v|^2 d\mu$$

(110) $$\leq C\sigma \int_{B_\sigma(x_0) \setminus \mathcal{S}_j(u)} \xi |\nabla v|^2 d\mu.$$

(Note that again we used the Lipschitz property of v in order to bound one term of $|\nabla v|$.) Next, we derive an estimate for $(iii)_2$ in a similar way as in $(III)_2$ and $(ii)_2$ to account for the difference in the C^1 estimates of $\Box(V,v)$ and $\mathbf{G}_{12}(V,v)$. We obtain

$$(111) \qquad (iii)_2 \leq C \int_{B_\sigma(x_0)} \xi d^2(v, P_0) + C\sigma \int_{B_\sigma(x_0)} |\nabla v|^2 d\mu.$$

Combining inequalities (109), (110) and (111) proves (104) and finishes the proof. $\square$

Lemma 52 implies the following analogue of the domain variation formula (2.3) of [**GS**].

PROPOSITION 53. *Let $u = (V,v) : B_{\sigma_*}(x_*) \to (\mathbf{R}^j \times Y_2^{k-j}, d_G)$ be a harmonic map satisfying the assumptions of Section 5. There exist $R_0 > 0$ and $C > 0$ such that for $x_0 \in \mathcal{S}_j(u) \cap B_{\frac{\sigma_*}{2}}(x_*)$ and $\sigma \in (0, R_0)$, we have*

$$(112) \qquad \frac{\frac{d}{d\sigma} E_{x_0}^v(\sigma)}{E_{x_0}^v(\sigma)} + \frac{2 - n + C\sigma}{\sigma} \geq \frac{2 \int_{\partial B_\sigma(x_0)} \left|\frac{\partial v}{\partial r}\right|^2 d\Sigma}{E_{x_0}^v(\sigma)}.$$

Furthermore, C depends only on the constant in the estimates (28)–(32) for the target metric G, the domain metric g and the Lipschitz constant of u.

PROOF. We will write $E = E_{x_0}^v$ and $I = I_{x_0}^v$ for simplicity. By Lemma 52,

$$-\frac{d}{dt}\bigg|_{t=0} E(\sigma) \leq C \int_{B_\sigma(x_0)} \xi d^2(v, P_0) d\mu + C\sigma \int_{B_\sigma(x_0)} \xi |\nabla v|^2 d\mu.$$

As in [**GS**] p.192-193, after letting ξ approximate the characteristic function, we obtain

$$(2 - n + C\sigma) E(\sigma) + \sigma \int_{\partial B_\sigma(x_0)} |\nabla v|^2 d\Sigma - 2\sigma \int_{\partial B_\sigma(x_0)} \left|\frac{\partial v}{\partial r}\right|^2 d\Sigma$$
$$\geq -C \int_{B_\sigma(x_0)} d^2(v, P_0) d\mu.$$

Combining the above with (77) and dividing by $\sigma E(\sigma)$, we obtain

$$\frac{E'(\sigma)}{E(\sigma)} + \frac{2 - n + C\sigma}{\sigma} \geq \frac{2 \int_{\partial B_\sigma(x_0)} \left|\frac{\partial v}{\partial r}\right|^2 d\Sigma}{E(\sigma)} - C\sigma \frac{I(\sigma)}{\sigma E(\sigma)}.$$

Proposition 43 asserts that there exists $R_0 > 0$ such that

$$-\sigma \frac{I(\sigma)}{\sigma E(\sigma)} \geq -2\sigma, \quad \forall \sigma \in (0, R_0).$$

The assertion immediately follows from combining the above two inequalities. $\square$

CHAPTER 9

Order Function

The main result of this section is to prove the following existence property of the order for the singular component of a harmonic map.

PROPOSITION 54. *Let $u = (V, v) : B_{\sigma_*}(x_*) \to (\mathbf{R}^j \times Y_2^{k-j}, d_G)$ be a harmonic map as in (16). For $x \in \mathcal{S}_j(u) \cap B_{\frac{\sigma_*}{2}}(x_*)$ and $0 < \sigma < \sigma_0 =: \sup\{\sigma : B_\sigma(x) \subset B_{\sigma_*}(x_*)\}$, assume that v is not constant in any neighborhood of x and define*

(113) $$Ord^v(x, \sigma) := \frac{\sigma \, E_x^v(\sigma)}{I_x^v(\sigma)}.$$

Then, there exist constants $C > 0$, $C_1 > 0$ and $R_0 > 0$ such that for any $x \in \mathcal{S}_j(u) \cap B_{\frac{\sigma_*}{2}}(x_*)$, there exists a function $\sigma \mapsto J_x(\sigma)$ with the properties

(114) $$e^{-C_1 \sigma} I_x^v(\sigma) \leq J_x(\sigma) \leq I_x^v(\sigma) e^{C_1 \sigma}, \quad \forall \sigma \in (0, R_0)$$

and

(115) $$\sigma \mapsto e^{C\sigma} \frac{\sigma \, E_x^v(\sigma)}{J_x(\sigma)} \text{ is non-decreasing in } (0, R_0).$$

Thus,

$$Ord^v(x) := \lim_{\sigma \to 0} Ord^v(x, \sigma)$$

exists and

(116) $$Ord^v(x) \leq e^{(C+C_1)\sigma} \frac{\sigma E_x^v(\sigma)}{I_x^v(\sigma)}, \quad \forall \sigma \in (0, R_0).$$

The constants C_1, C and R_0 can be chosen to depend continuously on x and to depend only on the constant in the estimates (28)-(32) for the target metric G, the domain metric g and the Lipschitz constant of u.

PROOF. Fix $x \in \mathcal{S}_j(u)$. For notational simplicity, let $I(\sigma) = I_x^v(\sigma)$ and $E(\sigma) = E_x^v(\sigma)$. Recall (cf. [**GS**] p.193) the equality

(117) $$\frac{I'(\sigma)}{I(\sigma)} = \frac{\int_{\partial B_\sigma(x)} \frac{\partial}{\partial r} d^2(v, P_0) d\Sigma}{I(\sigma)} + \frac{n - 1 + O(\sigma^2)}{\sigma}$$

where $O(\sigma)$ depends only on g. Combining (117) with (112), we obtain

(118) $$\frac{I'(\sigma)}{I(\sigma)} - \frac{E'(\sigma)}{E(\sigma)} - \frac{1}{\sigma}$$

$$\leq \frac{\left(E(\sigma) \int_{\partial B_\sigma(x)} \frac{\partial}{\partial r} d^2(v, P_0) d\Sigma - 2I(\sigma) \int_{\partial B_\sigma(x)} \left|\frac{\partial v}{\partial r}\right|^2 d\Sigma \right)}{E(\sigma) I(\sigma)} + C.$$

55

Now note that (117) implies
$$\int_{\partial B_\sigma(x)} \frac{\partial}{\partial r} d^2(v, P_0) d\Sigma \leq I'(\sigma)$$
for $\sigma > 0$ sufficiently small. Furthermore, Lemma 45 (cf. (77)) and Proposition 43 imply that

(119) $$\int_{B_\sigma(x)} d^2(v, P_0) d\mu \leq C(\sigma I(\sigma) + \sigma^2 E(\sigma)) \leq C\sigma^2 E(\sigma)$$

for $\sigma > 0$ sufficiently small. Thus, Proposition 40 implies that

$$E(\sigma) \int_{\partial B_\sigma(x)} \frac{\partial}{\partial r} d^2(v, P_0) d\Sigma - 2I(\sigma) \int_{\partial B_\sigma(x)} \left|\frac{\partial v}{\partial r}\right|^2 d\Sigma$$
$$\leq \frac{1}{2} \left(\int_{\partial B_\sigma(x)} \frac{\partial}{\partial r} d^2(v, P_0) d\Sigma + C \int_{B_\sigma(x)} d^2(v, P_0) d\mu \right)$$
$$\times \left(\int_{\partial B_\sigma(x)} \frac{\partial}{\partial r} d^2(v, P_0) d\Sigma \right) - 2I(\sigma) \int_{\partial B_\sigma(x)} \left|\frac{\partial v}{\partial r}\right|^2 d\Sigma$$
$$\leq 2I(\sigma) \int_{\partial B_\sigma(x)} \left|\frac{\partial}{\partial r} d(v, P_0)\right|^2 d\Sigma - 2I(\sigma) \int_{\partial B_\sigma(x)} \left|\frac{\partial v}{\partial r}\right|^2 d\Sigma.$$

(120) $$+C\sigma^2 E(\sigma) I'(\sigma).$$

Combining (118) with (120), we conclude that there exists $R_0 > 0$ such that

(121) $$0 \leq \frac{E'(\sigma)}{E(\sigma)} + \frac{1}{\sigma} - (1 - C\sigma^2)\frac{I'(\sigma)}{I(\sigma)} + C, \quad \text{for a.e. } \sigma \in (0, R_0).$$

Note that C and R_0 depend only on the constant in the estimates (28)–(32) for the target metric G, the domain metric g and the Lipschitz constant of u, and thus can be chosen to depend continuously on x.

Inequality (121) was first considered in [**Me**] formula (15) and subsequently in [**DM1**] formula (3.22). The existence of the limit follows as a special case of [**DM1**] Corollary 3.1. Note that since v is Lipschitz, we have by [**GS**] pp. 200–201 that in the definition of the order we can take $I(\sigma) = I(\sigma, v(0))$ instead of $I(\sigma, Q_\sigma)$. Therefore, if we set

$$J_x(\sigma) = I(\sigma) \exp\left(C \int_0^\sigma s^2 \frac{d}{ds} \log I(s) ds \right)$$

(note that the error terms in [**DM1**] are $O(\sigma)$ and not $O(\sigma^2)$, and this accounts for the difference in the definition of $J(\sigma)$), then (114) follows from [**DM1**] formula 3.32 and (115) follows from [**DM1**] Lemma 3.7. Inequality (116) follows immediately from (114) and (115). □

REMARK 55. The above Proposition works in great generality, and it implies that if a Lipschitz map satisfies the domain, the target variation formulas and the lower order bound, then it also satisfies the monotonicity formula (121) and has a well defined order. Formulas (114) - (116) follow as a formal consequence of (121).

Let $x_0 \in S_j(u) \cap B_{\frac{\sigma_*}{2}}(x_*)$ and assume $x_i = x_0$ for all i in (82) and that v is not constant in any neighborhood of x_0. Then Proposition 54 implies that the quantity

9. ORDER FUNCTION

$\frac{\sigma E_{x_0}^v(\sigma)}{I_{x_0}^v(\sigma)}$ is bounded above for $\sigma > 0$ small. Hence, there exists a sequence of blow up maps v_i converging to a harmonic map v_0 by Lemma 49.

DEFINITION 56. The harmonic map $v_0 : (B_1(0), \delta) \to Y_0$ above is called a *tangent map of v at x_0*.

LEMMA 57. *Let $u = (V, v) : B_{\sigma_*}(x_*) \to (\mathbf{R}^j \times Y_2^{k-j}, d_G)$ be a harmonic map as in (16). If v_0 is a tangent map of v at $x_0 \in \mathcal{S}_j(u) \cap B_{\frac{\sigma_*}{2}}(x_*)$, then v_0 is a homogeneous map and $Ord^{v_0}(0) = Ord^v(x_0)$.*

PROOF. Assume on the contrary that v_0 is not a homogeneous map. By [**GS**] Lemma 3.2 (by replacing the Riemannian simplicial complex by an arbitrary NPC target space) there exists $R \in (0, 1)$ sufficiently small such that

$$(122) \qquad Ord^{v_0}(x_0) < \frac{rE^{v_0}(r)}{I^{v_0}(r)}, \quad \forall r \in [R, 1].$$

For each σ_i, we choose $r_i \in (R, \frac{R+1}{2})$ such that

$$(123) \qquad \int_{\partial B_{r_i \sigma_i}(x_0)} |\nabla v|^2 d\Sigma \leq \frac{2}{(1-R)\sigma_i} \int_{B_{\sigma_i}(x_0)} |\nabla v|^2 d\mu \leq \frac{C}{\sigma_i} E^v(\sigma_i).$$

Here and henceforth, C will denote an arbitrary constant that is independent of i. Now note that the map v is not a competitor of the harmonic map $_{\sigma_i} w$ in the domain $B_{r_i \sigma_i}(x_0)$ because $_{\sigma_i} w$ does not necessarily agree with v on $\partial B_{r_i \sigma_i}(x_0)$. Therefore, we "bridge" the gap between v and $_{\sigma_i} w$ using [**KS2**] Lemma 3.12 to define a map $_{\sigma_i} \bar{w}$ with the same boundary value as v. More precisely, for $\rho > 0$ small, we let $F : B_{r_i \sigma_i - \rho}(x_0) \to B_{r_i \sigma_i}(x_0)$ be the scaling map $F(x) = x_0 + \frac{r_i \sigma_i}{r_i \sigma_i - \rho}(x - x_0)$ and set

$$\bar{v}(x) = \begin{cases} v \circ F(x) & \text{for } x \in B_{r_i \sigma_i - \rho}(x_0), \\ W(x) & \text{for } x \in B_{r_i \sigma_i}(x_0) \setminus B_{r_i \sigma_i - \rho}(x_0) \end{cases}$$

where

$$(124) \qquad W : B_{r_i \sigma_i}(x_0) \setminus B_{r_i \sigma_i - \rho}(x_0) \simeq \partial B_{r_i \sigma_i}(x_0) \times [0, \rho] \to Y_2^{k-j}$$

is the interpolation map between $_{\sigma_i} w \big|_{\partial B_{r_i \sigma_i}(x_0)}$ and $v \big|_{\partial B_{r_i \sigma_i}(x_0)}$

$$W(y, s) = (1 - \frac{s}{\rho}) v(y) + \frac{s}{\rho} \, _{\sigma_i} w(y).$$

Thus, $W = v \circ F$ on $\partial B_{r_i \sigma_i - \rho}(x_0)$ and $W = \, _{\sigma_i} w$ on $\partial B_{r_i \sigma_i}(x_0)$. The energy of $\bar{v}$ is close to that of v inside the ball $B_{r_i \sigma_i}(x_0)$; more precisely, since $\bar{v}\big|_{B_{r_i \sigma_i - \rho}(x_0)}$ and $v\big|_{B_{r_i \sigma_i}(x_0)}$ differ only by scaling, we can bound the difference by

$$(125) \quad E^{\bar{v}}(r_i \sigma_i) - E^v(r_i \sigma_i)$$
$$\leq \left(\frac{r_i \sigma_i}{r_i \sigma_i - \rho}\right)^2 E^v(r_i \sigma_i) - E^v(r_i \sigma_i) + E^W \leq \frac{C\rho}{\sigma_i} E^v(r_i \sigma_i) + E^W$$

provided ρ small compared to σ_i (in fact, later we set $\rho = \sigma_i^2$). Furthermore, by [**KS2**] (3.23)

$$(126) \qquad E^W \leq \frac{C\rho}{2} \int_{\partial B_{r_i \sigma_i}(x_0)} |\nabla v|^2 + |\nabla_{\sigma_i} w|^2 d\Sigma + \frac{C}{\rho} \int_{\partial B_{r_i \sigma_i}(x_0)} d^2(v, \, _{\sigma_i} w) d\Sigma$$

(The constant C comes from the fact that the metric in the annulus does not correspond with the product metric via (124)). By the Lipschitz estimate [**KS1**] Theorem 2.4.6 applied to $_{\sigma_i}w$, we obtain

$$(127) \qquad |\nabla_{\sigma_i} w|^2 \leq \frac{C}{\sigma_i^n} E^v(\sigma_i) \text{ in } B_{\frac{1+R}{2}}(x_0).$$

Applying (123) and (127) in (126), we obtain

$$(128) \qquad E^W \leq \frac{C\rho}{\sigma_i} E^v(\sigma_i) + \frac{C}{\rho} \int_{\partial B_{r_i \sigma_i}(x_0)} d^2(v, {_{\sigma_i}w}) d\Sigma.$$

The fact that $\bar{v}$ is a competitor for $_{\sigma_i}w$, (125) and (128) imply

$$\begin{aligned} E^{\sigma_i w}(r_i \sigma_i) &- E^v(r_i \sigma_i) \\ &\leq E^{\sigma_i w}(r_i \sigma_i) - E^{\bar{v}}(r_i \sigma_i) + E^{\bar{v}}(r_i \sigma_i) - E^v(r_i \sigma_i) \\ &\leq \frac{C\rho}{\sigma_i} E^v(\sigma_i) + C E^W \\ &\leq \frac{C\rho}{\sigma_i} E^v(\sigma_i) + \frac{C}{\rho} \int_{\partial B_{r_i \sigma_i}(x_0)} d^2(v, {_{\sigma_i}w}) d\Sigma. \end{aligned}$$

Thus, by rescaling and applying Proposition 48 and the uniform bound $E^{v_{\sigma_i}}(1) \leq 2\alpha$, we obtain

$$\begin{aligned} E^{w_{\sigma_i}}(r_i) - E^{v_{\sigma_i}}(r_i) &\leq \frac{C\rho}{\sigma_i} E^{v_{\sigma_i}}(1) + \frac{C\sigma_i}{\rho} \int_{\partial B_{r_i}(x_0)} d^2(v_{\sigma_i}, w_{\sigma_i}) d\Sigma \\ &\leq \frac{C\rho}{\sigma_i} + \frac{C\sigma_i^3}{\rho}. \end{aligned}$$

Thus, by choosing $\rho = \sigma_i^2$, we have

$$(129) \qquad E^{w_{\sigma_i}}(r_i) - E^{v_{\sigma_i}}(r_i) \leq C\sigma_i.$$

We can similarly define

$$_{\sigma_i}\bar{w}(x) = \begin{cases} {_{\sigma_i}w} \circ F(x) & \text{for } x \in B_{r_i \sigma_i - \rho}(x_0), \\ \overline{W}(x) & \text{for } x \in B_{r_i \sigma_i}(x_0) \setminus B_{r_i \sigma_i - \rho}(x_0) \end{cases}$$

where $\overline{W}$ is the interpolation map between $_{\sigma_i}w$ and v so that $\overline{W} = {_{\sigma_i}w} \circ F$ on $\partial B_{r_i \sigma_i - \rho}(x_0)$ and $\overline{W} = v$ on $\partial B_{r_i \sigma_i}(x_0)$. The energy of $\hat{u} = (V, {_{\sigma_i}w})$ is close to that of $\bar{u} = (V, {_{\sigma_i}\bar{w}})$ inside the ball $B_{r_i \sigma_i}$; more precisely, we can bound the difference using Lemma 29 by

$$\begin{aligned} E^{\bar{u}}(r_i \sigma_i) &- E^{\hat{u}}(r_i \sigma_i) \\ &\leq \left(\frac{r_i \sigma_i}{r_i \sigma_i - \rho}\right)^2 E^{\sigma_i w}(r_i \sigma_i) - E^{\sigma_i w}(r_i \sigma_i) + E^{\overline{W}} \\ &\quad + C \int_{B_{r_i \sigma_i}(x_0)} d^2(v, P_0) + d^2({_{\sigma_i}w}, P_0) d\mu \\ &\leq \frac{C\rho}{\sigma_i} E^{\sigma_i w}(r_i \sigma_i) + E^{\overline{W}} + C \int_{B_{r_i \sigma_i}(x_0)} d^2(v, P_0) + d^2({_{\sigma_i}w}, P_0) d\mu. \end{aligned}$$

Integrating inequality (78) over $B_{r_i\sigma_i}(x_0)$ and using the fact that $\bar{u}$ is a competitor for the harmonic map u, we obtain

(130)
$$\begin{aligned} E^v(r_i\sigma_i) &- E^{\sigma_i w}(r_i\sigma_i) \\ &\leq E^u(r_i\sigma_i) - E^{\hat{u}}(r_i\sigma_i) + C\int_{B_{r_i\sigma_i}(x_0)} d^2(v,P_0) + d^2(\sigma_i w, P_0) d\mu \\ &\leq E^u(r_i\sigma_i) - E^{\bar{u}}(r_i\sigma_i) + E^{\bar{u}}(r_i\sigma_i) - E^{\hat{u}}(r_i\sigma_i) \\ &\quad + C\int_{B_{r_i\sigma_i}(x_0)} d^2(v,P_0) + d^2(\sigma_i w, P_0) d\mu \\ &\leq \frac{C\rho}{\sigma_i} E^{\sigma_i w}(r_i\sigma_i) + CE^{\overline{W}} + C\int_{B_{r_i\sigma_i}(x_0)} d^2(v,P_0) + d^2(\sigma_i w, P_0) d\mu. \end{aligned}$$

We can bound $E^{\overline{W}}$ in an analogous way as E^W, hence by scaling, applying Lemma 45 and Lemma 46, noting that $E^{w_{\sigma_i}}(1) \leq E^{v_{\sigma_i}}(1) \leq 2\alpha$ and letting $\rho = \sigma_i^2$, we obtain

$$\begin{aligned} E^{v_{\sigma_i}}(r_i) &- E^{w_{\sigma_i}}(r_i) \\ &\leq \frac{C\rho}{\sigma_i} + \frac{2C\sigma_i^3}{\rho} + C\int_{B_{\sigma_i}(x_0)} d^2(v,P_0) + d^2(\sigma_i w, P_0) d\mu \\ &\leq \frac{C\rho}{\sigma_i} + \frac{2C\sigma_i^3}{\rho} + C\sigma_i \end{aligned}$$

(131)
$$\leq C\sigma_i.$$

Combining (129) and (131),

(132)
$$|E^{v_{\sigma_i}}(r_i) - E^{w_{\sigma_i}}(r)| \leq C\sigma_i,$$

and we can deduce

(133)
$$\frac{r_i(E^{v_{\sigma_i}}(r_i) - C\sigma_i)}{I^{w_{\sigma_i}}(r_i)} \leq \frac{r_i E^{w_{\sigma_i}}(r_i)}{I^{w_{\sigma_i}}(r_i)} \leq \frac{r_i(E^{v_{\sigma_i}}(r_i) + C\sigma_i)}{I^{w_{\sigma_i}}(r_i)}.$$

By taking a subsequence if necessary, we can assume $r_i \to r_0 \in [R, \frac{R+1}{2}]$. Recall that w_{σ_i} is a sequence of harmonic maps with uniformly bounded Lipschitz constant in $B_{\frac{R+1}{2}}(0)$ (cf. (127)). Thus, $I^{w_{\sigma_i}}(r_i) \to I^{v_0}(r_0)$ and $E^{w_{\sigma_i}}(r_i) \to E^{v_0}(r_0)$ by [**KS2**] Proposition 3.7 and Theorem 3.11. Furthermore, $I^{v_{\sigma_i}}(r_i) \to I^{v_0}(r_0)$ by Proposition 48. Therefore

$$\begin{aligned} \lim_{i\to\infty} \frac{r_i(E^{v_{\sigma_i}}(r_i) \pm C\sigma_i)}{I^{w_{\sigma_i}}(r_i)} &= \lim_{i\to\infty}\left(\frac{I^{v_{\sigma_i}}(r_i)}{I^{w_{\sigma_i}}(r_i)}\frac{r_i E^{v_{\sigma_i}}(r_i)}{I^{v_{\sigma_i}}(r_i)} \pm \frac{Cr_i\sigma_i}{I^{w_{\sigma_i}}(r_i)}\right) \\ &= \lim_{i\to\infty} \frac{r_i E^{v_{\sigma_i}}(r_i)}{I^{v_{\sigma_i}}(r_i)} \\ &= \lim_{i\to\infty} \frac{r_i\sigma_i E^v(r_i\sigma_i)}{I^v(r_i\sigma_i)} \\ &= Ord^v(x_0), \end{aligned}$$

and we conclude by taking limits as $i \to \infty$ of (133) that

(134)
$$Ord^v(x_0) = \frac{r_0 E^{v_0}(r_0)}{I^{v_0}(r_0)}.$$

This contradicts (122), thereby proving that v_0 is a homogeneous map. $\square$

The following are Corollaries of Proposition 54.

COROLLARY 58. *Let $u = (V, v) : B_{\sigma_*}(x_*) \to (\mathbf{R}^j \times Y_2^{k-j}, d_G)$ be a harmonic map as in (16). If $v \equiv P_0$ on any open subset of $B_{\frac{\sigma_*}{2}}(x_*)$, then $v \equiv P_0$ in $B_{\frac{\sigma_*}{2}}(x_*)$.*

PROOF. If v is not constant in $B_{\frac{\sigma_*}{2}}(x_*)$ but identically equal to P_0 on an open subset of $B_{\frac{\sigma_*}{2}}(x_*)$, then there exists a ball $B \subset B_{\frac{\sigma_*}{2}}(x_*)$ such that $v \equiv P_0$ in the interior of B, but for some $x_0 \in \partial B$, v is not constant in any neighborhood of x_0. Let $v_0 : B_1(0) \to Y_0$ be the tangent map of v at x_0. Then v_0 is identically constant on half of $B_1(0)$ and this contradicts Proposition 3.4 of [**GS**]. □

COROLLARY 59. *Let $u = (V, v) : B_{\sigma_*}(x_*) \to (\mathbf{R}^j \times Y_2^{k-j}, d_G)$ be a harmonic map as in (16). Then, there exists $A > 0$ such that for $x \in \mathcal{S}_j(u) \cap B_{\frac{\sigma_*}{2}}(x_*)$, we have*
$$Ord^v(x) \leq A.$$

PROOF. Since
$$\int_\sigma^{\sigma_0} s \frac{d}{ds} \log I_x^v(s) ds = \sigma_0 \log I_x^v(\sigma_0) - \sigma \log I_x^v(\sigma) - \int_\sigma^{\sigma_0} \log I_x^v(s) ds,$$
the map $x \mapsto J_x(\sigma)$ is a continuous map and $J_x(\sigma) \neq 0$. Thus the map $x \mapsto \frac{\sigma E_x^v(\sigma)}{J_x(\sigma)}$ is continuous, and the result follows from the fact that a non-increasing limit of continuous functions is upper semicontinuous. □

COROLLARY 60. *Let $u = (V, v) : B_{\sigma_*}(x_*) \to (\mathbf{R}^j \times Y_2^{k-j}, d_G)$ be a harmonic map as in (16). Then there exist $C > 0$ and $R_0 > 0$ such that for any $x \in \mathcal{S}_j(u) \cap B_{\frac{\sigma_*}{2}}(x_*)$, we have*
$$\sigma \mapsto e^{C\sigma} \frac{I_x^v(\sigma)}{\sigma^{n-1+2\alpha}} \quad \text{and} \quad \sigma \mapsto e^{C\sigma} \frac{E_x^v(\sigma)}{\sigma^{n-2+2\alpha}}$$
are monotone non-decreasing in $(0, R_0)$. The constants C_1, C and R_0 can be chosen to depend continuously on x and depend only on the constant in the estimates (28)–(32) for the target metric G, the domain metric g and the Lipschitz constant of u.

PROOF. Let $I(\sigma) = I_x^v(\sigma)$, $E(\sigma) = E_x^v(\sigma)$ and $J(\sigma) = J_x(\sigma)$. Combining Proposition 40 with (119) and Corollary 59, we obtain
$$2E(\sigma) \leq \int_{\partial B_\sigma(x)} \frac{\partial}{\partial r} d^2(v, P_0) d\mu + \sigma I(\sigma) + C\sigma E(\sigma)$$
$$\leq I'(\sigma) - \frac{n-1}{\sigma} I(\sigma) + CI(\sigma).$$
Since Proposition 54 implies
$$e^{-C\sigma} \alpha I(\sigma) \leq e^{-C\sigma} \alpha J(\sigma) \leq \sigma E(\sigma), \quad \forall \sigma \in (0, R_0),$$
we obtain
$$2\alpha I(\sigma) \leq \sigma I'(\sigma) - (n-1) I(\sigma) + C\sigma I(\sigma), \quad \forall \sigma \in (0, R_0).$$
In the above the constant C depends as before on the constant in the estimates (28)-(32) for the target metric G, the domain metric g and the Lipschitz constant of u. By rearranging, we obtain
$$\frac{d}{d\sigma} \log \left(\frac{I(\sigma)}{\sigma^{n-1+2}} \right) = \frac{I'(\sigma)}{I(\sigma)} - \frac{n-1+2\alpha}{\sigma} \geq -C, \quad \forall \sigma \in (0, R_0)$$

Combining this with inequality (121), we obtain
$$\frac{d}{d\sigma}\log\left(\frac{E(\sigma)}{\sigma^{n-2+2}}\right) = \frac{E'(\sigma)}{E(\sigma)} - \frac{n-2+2\alpha}{\sigma} \geq -C, \quad \forall \sigma \in (0, R_0).$$
The above two inequalities immediately imply the assertion of the Corollary. $\square$

CHAPTER 10

The Gap Theorem

First recall the ϵ-gap Theorem 6.3 of [**GS**] which states that if X is a F-connected complex and K a bounded subset of X, then there exists $\epsilon_0 > 0$ such that for any harmonic map $u : (B_1(0), g) \to X$ with $u(B_1(0)) \subset K$, either

(135) $$Ord^u(0) = 1 \text{ or } Ord^u(0) \geq 1 + \epsilon_0.$$

This gap property also holds for a DM-complex.

THEOREM 61. *If (Y, d_G) is a NPC DM-complex, K is a bounded subset of Y, there exists $\epsilon_0 > 0$ depending only on K and n such that for any harmonic map $u : (B_1(0), g) \to (Y, d_G)$ with $u(0) \subset K$,*

$$Ord^u(0) = 1 \text{ or } Ord^u(0) \geq 1 + \epsilon_0.$$

PROOF. On the contrary, assume there exists a sequence of harmonic maps $\{u_i\}$ with $u_i(0) \subset K$ and

(136) $$1 < Ord^{u_i}(0) < 1 + \frac{1}{i}.$$

Let $u_{i\sigma}$ be the σ-blow up map of u_i. By the monotonicity properties of u, we can choose $\sigma_i \to 0$ such that

$$E^{u_{i\sigma_i}}(1) < 1 + \frac{1}{i} \text{ and } I^{u_{i\sigma_i}}(1) = 1.$$

Since $\overline{K}$ is compact, there is only a finite number of homeomorphism types that appear as tangent cones at $P \in K$. Hence, we can assume that $u_{i\sigma_i}$ maps into a single cone, i.e.

$$u_{i\sigma_i} = (V_i, v_i) : B_1(0) \to (\mathbf{R}^j \times Y_2^{k-j}, G_i).$$

Here, the metric G_i is the appropriate blow up metric at $u_i(0)$ as in (9). We may also assume (by taking a subsequence if necessary) that $u_i(0) \to Q_0 \in \overline{K}$. Since G is a smooth metric up to its boundary on each simplex and $\sigma_i \to 0$, G_i converges smoothly to a Euclidean metric G_0. Finally, we may assume that j is the maximal integer such that $u_{i\sigma_i}$ can be represented in the above form; i.e. there does not exist $j' > j$ and $\sigma \in (0, 1]$ such that $u_{i\sigma_i}\big|_{B_\sigma(0)}$ maps into a cone $\mathbf{R}^{j'} \times Z^{k-j'}$. Let $u_{i*} = (V_{i*}, v_{i*})$ be a tangent map of u_i at 0. Here, $V_{i*} : B_1(0) \to \mathbf{R}^j$ is a harmonic map into Euclidean space. Since $1 < Ord^{u_{i*}}(0) = Ord^{u_i}(0) < 1 + \frac{1}{i}$, we conclude that $V_{i*} \equiv 0$.

The maps $\{u_{i\sigma_i}\}$ are uniformly Lipschitz with respect to G_0 and the energy of $u_{i\sigma_i}$ with respect to G_0 is within ϵ_i of minimizing where $\epsilon_i \to 0$ as $i \to \infty$. Thus, (after taking a subsequence if necessary) we can assume that $u_{i\sigma_i}$ converges locally uniformly to a non-constant harmonic map $u_0 = (V_0, v_0) : B_1(0) \to (\mathbf{R}^j \times Y_2^{k-j}, G_0)$

and the energy of $u_{i\sigma_i}|_{B_r(0)}$ converges to that of $u_0|_{B_r(0)}$ for all $r \in (0,1)$ (cf. [**KS2**] Theorem 3.11). Thus,

$$\frac{rE^{u_0}(r)}{I^{u_0}(r)} = \lim_{i \to \infty} \frac{rE^{u_{i\sigma_i}}(r)}{I^{u_{i\sigma_i}}(r)} = 1, \ \forall r \in (0,1).$$

This implies that $u_0 = (V_0, v_0)$ is a homogeneous map of degree 1 (cf. [**GS**] Lemma 3.2). We claim that v_0 is a constant map. Indeed, if v_0 is not a constant, then v_0 is effectively contained a subcomplex $\mathbf{R}^l \times Y_3^{k-j-l}$ of $\mathbf{R}^j \times Y_2^{k-j}$ (cf. [**GS**] Proposition 3.1 and Lemma 6.2). By [**GS**] Theorem 5.1, there exists $r_0 > 0$ such that $u_{i\sigma_i}(B_{r_0}(0)) \subset \mathbf{R}^{j+l} \times Y_3^{k-j-l}$ for i sufficiently large. This contradicts the maximality of j proving the claim. Since v_0 is a constant map, V_0 is a non-constant map. The proof of Lemma 50 implies that the $C^{1,\beta}$ norm of V_i is uniformly bounded in $B_{\frac{1}{2}}(0)$. Hence (by Arzela-Ascoli and taking a subsequence if necessary), we may assume that $\frac{\partial V_{i*}}{\partial x^\alpha}$ converges to $\frac{\partial V_0}{\partial x^\alpha}$. Thus, V_{i*} is not a constant map for sufficiently large i, a contradiction to the conclusion in the previous paragraph. □

As a consequence of Theorem 61, we have the following

PROPOSITION 62. *If $u : (\Omega, g) \to (Y, d_G)$ is a harmonic map from a Riemannian domain into a DM-complex and $u = (V, v) : (B_{\sigma_*}(x_*), g) \to (\mathbf{R}^j \times Y_2^{k-j}, d_G)$ a local representation as in (16), then there exists $\epsilon_0 > 0$ such that*

$$Ord^u(x_0) \geq 1 + \epsilon_0, \ \forall x_0 \in \mathcal{S}_0(u) \cap B_{\frac{\sigma_*}{2}}(x_*)$$

and

$$\dim_\mathcal{H}\left(\mathcal{S}_0(u) \cap B_{\frac{\sigma_*}{2}}(x_*)\right) \leq n - 2.$$

PROOF. By the interior Lipschitz continuity of u, we can choose a bounded set K such that $u(B_{\frac{\sigma_*}{2}}(x_*)) \subset K$. The first assertion follows from Theorem 61. A tangent map u_* of u maps into an F-connected complex, so $\dim(\mathcal{S}_0(u_*)) \leq n-2$ by [**GS**] Theorem 6.4. Combining this with the first assertion, we can apply Theorem 78 of Appendix 2 with $\mathcal{S} := \mathcal{S}_0(u) \cap B_{\frac{\sigma_*}{2}}(x_*)$ to prove the second assertion. □

Additionally, we need an analogous statement for the singular component map.

PROPOSITION 63. *Under the same assumptions as in Proposition 62 and under the assumptions of Section 5, there exists $\epsilon_0 > 0$ such that*

$$Ord^v(x_0) \geq 1 + \epsilon_0, \ \forall x_0 \in \mathcal{S}_j(u) \cap B_{\frac{\sigma_*}{2}}(x_*)$$

and

$$\dim_\mathcal{H}\left(\mathcal{S}_j(u) \cap B_{\frac{\sigma_*}{2}}(x_*)\right) \leq n - 2.$$

PROOF. As before choose a bounded set K such that $u(B_{\frac{\sigma_*}{2}}(x_*)) \subset K$. The proof closely follows that of Theorem 61, and we assume to the contrary that there exists a sequence of points $x_i \in \mathcal{S}_j(u) \cap B_{\frac{\sigma_*}{2}}(x_*)$ such that

$$1 < Ord^v(x_i) < 1 + \frac{1}{i}.$$

On the other hand, the proof here differs from that of Theorem 61 in that instead of using a σ_i-blow up map of u_i (as done in Theorem 61), we use the σ_i-blow up map $v_i := v_{\sigma_i, x_i}$ and the σ_i-approximate harmonic blow up map $w_i = w_{\sigma_i, x_i}$ of v at x_i (cf. Definition 44). Indeed, by Proposition 54, we can choose $\sigma_i \to 0$ such that

$$E^{w_i}(1) \leq E^{v_i}(1) < 1 + \frac{1}{i} \text{ and } , \int_{\partial B_1(0)} d^2(w_i, P_0) d\Sigma_i = 1.$$

We can thus argue as in the proof of Theorem 61 to obtain a homogeneous degree 1 harmonic map $v_0 : B_1(0) \to (Y_2^{k-j}, d_h)$ into a F-connected complex as a limit (under uniform convergence on compact sets) of the sequence $\{w_i\}$, and hence of $\{v_i\}$ (by Proposition 48). Furthermore, the space (Y_2^{k-j}, d_h) is essentially regular by [**GS**] Theorem 6.3. Therefore, if $Ord^{v_0}(0) = 1$ then applying Proposition 67 of Appendix 1 with $l = v_0$, we conclude that for any i sufficiently large

$$\sup_{B_s(0)} d(v_i, P_0) > \lambda s \text{ for } s > 0 \text{ sufficiently small.}$$

Fix $i > 0$ sufficiently large and identify $x_i = 0$. We then have

$$\sup_{B_s(0)} d(v(\sigma_i x), P_0) > \lambda \mu_{\sigma_i} s \text{ for } s > 0 \text{ sufficiently small.}$$

By the monotonicity property of the harmonic map u, we then have for $\sigma > 0$ sufficiently small,

$$\mu_{\sigma\sigma_i}^{-1} d(v(\sigma\sigma_i x), P_0) > \mu_{\sigma\sigma_i}^{-1} \lambda \mu_{\sigma_i} \sigma$$
$$= \lambda \frac{\mu_{\sigma_i}}{\sigma_i} \sqrt{\frac{(\sigma\sigma_i)^{n+1}}{I^u(\sigma\sigma_i)}}$$
$$\geq \lambda \frac{\mu_{\sigma_i}}{\sigma_i} \sqrt{\frac{1}{e^c I^u(1)}}.$$

Thus, there exists a tangent map $u_* = (V_*, v_*)$ of u at x_i and a sequence $\sigma_l \to 0$ such that by replacing σ by σ_l in the above inequality, we obtain

$$d(v_*(x), P_0) \geq \lambda \frac{\mu_{\sigma_i}}{\sigma_i} \sqrt{\frac{1}{e^c I^u(1)}} > 0$$

which contradicts Lemma 20 and the fact that $x_i \in \mathcal{S}_j(u)$. We can thus conclude that there exists $\epsilon_0 > 0$ such that $Ord^v(x_0) \geq 1 + \epsilon_0$ for $x_0 \in \mathcal{S}_j(u) \cap B_{\frac{\sigma_*}{2}}(x_*)$.

For the second assertion, let $\mathcal{S} := \mathcal{S}_j(u) \cap B_{\frac{\sigma_*}{2}}(x_*)$. The map v and the set $\mathcal{S}$ satisfy Properties (P1)-(P3) of Appendix 2. Indeed, Proposition 54 implies that (P1) holds. Property (P2) asserts the existence of blow-up maps and approximating harmonic blow up maps for v as in Definition 44 and the required convergence $d(v_{\sigma_i}, w_{\sigma_i}) \to 0$ is established in Proposition 48. Moreover, as required in (P3), formula (132) holds. Proposition 63 implies that the order gap property in Appendix 2 is satisfied. Since a tangent map v_0 is a harmonic map into an F-connected complex, [**GS**] Theorem 6.4 implies that v satisfies the codimension 2 property of the tangent map with respect to $\mathcal{S}$ as in Theorem 78 of Appendix 2. Thus, the first assertion and Theorem 78 implies $\dim_{\mathcal{H}} \left(\mathcal{S}_j(u) \cap B_{\frac{\sigma_*}{2}}(x_*) \right) \leq n - 2$. □

CHAPTER 11

Proof of Theorems 1–4

We now turn to the proof of Theorem 1. **Fix a** $j \in \{k_0, \ldots, 1\}$ and let
$$u = (V, v) : (B_{\sigma_*}(x_*), g) \to (\mathbf{R}^j \times Y_2^{k-j}, d_G)$$
be a local representation of a harmonic map into a DM-complex (cf. (16)). Define the following:

STATEMENT 1[j]: $\dim_{\mathcal{H}}\left(\mathcal{S}(u) \cap B_{\frac{\sigma_*}{2}}(x_*)\right) \leq n - 2$.

STATEMENT 2[j]: For $q \in [1, 2)$ sufficiently close to 2 and any compactly contained subdomain Ω of $B_{\frac{\sigma_*}{2}}(x_*)$, there exists a sequence of smoth functions $\{\psi_i\}$ with $\psi_i \equiv 0$ in a neighborhood of $\mathcal{S}(u) \cap \overline{\Omega}$, $0 \leq \psi_i \leq 1$, $\psi_i \to 1$ for all $x \in \Omega \backslash \mathcal{S}(u)$ such that

(137) $$\lim_{i \to \infty} \int_{B_{\frac{\sigma_*}{2}}(x_*)} |\nabla u||\nabla \psi_i| \, d\mu = 0$$

(138) $$\lim_{i \to \infty} \int_{B_{\frac{\sigma_*}{2}}(x_*)} |\nabla u||\nabla \psi_i|^q \, d\mu = 0$$

and

(139) $$\lim_{i \to \infty} \int_{B_{\frac{\sigma_*}{2}}(x_*)} |\nabla \nabla u||\nabla \psi_i| \, d\mu = 0.$$

Our strategy is to prove STATEMENT 1[j] for all $j \in \{k_0 + 1, \ldots, 1\}$ which immediately proves Theorem 1. Similarly STATEMENT 2[j] for all $j \in \{k_0 + 1, \ldots, 1\}$ proves Theorem 2. We proceed with backwards induction on j. In order to use the results of the previous sections, we have to satisfy all the Assumptions of Chapter 5 and thus we have to prove both statements at the same time. The initial step is the case when $j = k_0 + 1$. Since $\mathcal{S}_{k_0+1}(u) = \emptyset$, Proposition 62 immediately implies STATEMENT 1[$k_0 + 1$]. Furthermore, using order gap property for u asserted in Proposition 62, we can apply the same proof as in [**GS**] Lemma 6.4 (with $\mathcal{S}$ replaced by $\mathcal{S}_0(u)$) to prove STATEMENT 2[$k_0 + 1$].

For the inductive step when $j \in \{k_0, \ldots, 1\}$, we assume that STATEMENT 1[$j+1$] and STATEMENT 2[$j + 1$] hold. Now, the assumptions of Chapter 5 are always satisfied except Assumption 3 (*ii*) and Assumption 6. However, by combining STATEMENT 1[$j + 1$] and Proposition 62 we obtain that Assumption 3 (*ii*) holds. Furthermore, by combining STATEMENT 2[$j + 1$] and a partition of unity argument Assumption 6 also holds. Under these assumptions, we now verify STATEMENT 1[j] and STATEMENT 2[j].

Proof of Statement 1[j]. Proposition 62, Proposition 63 and STATEMENT 1[$j+1$] immediately imply STATEMENT 1[j]. □

Proof of Statement 2[j]. Let $\epsilon_0 > 0$ be smaller than either of the ϵ_0 that appears in Proposition 62. Choose constants $q < 2$, $p > 2$, $\delta > 0$ and $D > 0$ satisfying the properties that $q < 2$ is sufficiently close to 2 such that the assertion of Assumption 6 holds and such that

$$\frac{1}{p} + \frac{1}{q} = 1, \quad D < \delta < \epsilon_0,$$

(140) $\qquad -2 + D < -q - q\delta \quad \text{and} \quad -2 + D < -p - p\delta + \epsilon_0.$

Let Ω be a subdomain compactly contained in $B_{\frac{\sigma_\star}{2}}(x_\star)$ and let Ω_2 be such that $\Omega \subset\subset \Omega_2 \subset\subset B_{\frac{\sigma_\star}{2}}(x_\star)$. Proposition 63 implies that $|\nabla v|(x) = 0$ for $x \in \mathcal{S}_j(u)$. Since any point in $\mathcal{S}_j(u)$ is of order 1, $|\nabla u|(x) \neq 0$ for $x \in \mathcal{S}_j(u)$. Hence, $|\nabla V|(x) \neq 0$ for $x \in \mathcal{S}_j(u)$. Since ∇V is Hölder continuous (by Lemma 50 and Sobolev embedding), this implies that there exists a neighborhood $\mathcal{N} \subset \Omega_2$ of $\mathcal{S}_j(u) \cap \overline{\Omega}$ and a constant λ_0 such that

(141) $\qquad |\nabla V| \geq \lambda_0 > 0 \text{ on } \mathcal{N}.$

Below, we will use C to denote any generic constant that depends only on λ_0, the dimension of n of the domain, the Lipschitz constant of u in Ω_2 and the $W^{1,p}$ norm of V. $\square$

Fix $i \in \mathbf{N}$. STATEMENT 1[j] implies that we can choose a finite covering $\{B_{r_J}(x_J) : J = 1, \ldots, l\}$ of the compact set $\mathcal{S}_j(u) \cap \overline{\Omega}$ satisfying

(142) $$\sum_{J=1}^l r_J^{n-2+D} < \frac{1}{i}.$$

By choosing $x_J \in \mathcal{S}_j(u) \cap \overline{\Omega}$ and r_J's sufficiently small, we can also assume

(143) $\qquad B_{3r_J}(x_J) \subset \mathcal{N}.$

Let φ_J be a smooth function such that $\varphi_J \equiv 0$ on $B_{r_J}(x_J)$, $\varphi_J \equiv 1$ on $\Omega \backslash B_{2r_J}(x_J)$, $|\nabla \varphi_J| \leq Cr_J^{-1}$ and $|\nabla\nabla \varphi_J| \leq Cr_J^{-2}$. Define φ by setting

$$\varphi = \prod_{J=1}^l \varphi_J.$$

Thus, $\varphi \equiv 0$ in a neighborhood of $\mathcal{S}_j(u) \cap \overline{\Omega}$, $\varphi \equiv 1$ outside $\bigcup_{J=1}^l B_{2r_J}(x_J)$ and $0 \leq \varphi \leq 1$. Let

$$\Omega_0 := \Omega \backslash \bigcup_{J=1}^l \overline{B_{r_J}(x_J)}.$$

We also define ρ_J to be a smooth function that is identically one on $B_{2r_J}(x_J)$ and identically zero on $\Omega \backslash B_{3r_J}(x_J)$ with $|\nabla \rho_J| \leq Cr_J^{-1}$, $|\nabla\nabla \rho_J| \leq Cr_J^{-2}$ and set

$$\rho = 1 - \prod_{J=1}^l \rho_J.$$

Since φ and ρ are now fixed, Proposition 62 implies that we can choose a finite covering $\{B_{s_J}(\xi_J) : J = 1, \ldots, l'\}$ of $\mathcal{S}_0(u) \cap v^{-1}(P_0) \cap \overline{\Omega}_0$ with

(144) $\qquad f(\sup_\Omega |\nabla \varphi|, \sup_\Omega |\nabla\nabla \varphi|, \sup_\Omega |\nabla \rho|) \sum_{J=1}^{l'} s_J^{n-2+D} < \frac{1}{i},$

where f is a certain function to be determined later.

11. PROOF OF THEOREMS 1-4

Let ϕ_J be a smooth function such that $\phi_J \equiv 0$ on $B_{s_J}(\xi_J)$, $\phi_J \equiv 1$ on $\Omega_0 \setminus B_{2s_J}(\xi_J)$, $|\nabla \phi_J| \leq C s_J^{-1}$ and $|\nabla \nabla \phi_J| \leq C s_J^{-2}$. Define ϕ by setting

$$\phi = \prod_{J=1}^{l'} \phi_J.$$

Thus, $\phi \equiv 0$ in a neighborhood of $\mathcal{S}_0(u) \cap v^{-1}(P_0) \cap \overline{\Omega}_0$, $\phi \equiv 1$ outside $\bigcup_{J=1}^{l'} B_{2s_J}(\xi_J)$ and $0 \leq \phi \leq 1$. Let

$$\Omega_1 := \Omega_0 \setminus \bigcup_{J=1}^{l'} \overline{B_{s_J}(\xi_J)}.$$

Let $\{\hat{\psi}_i\}$ be defined as $\{\psi_i\}$ in Assumption 6. By taking a subsequence if necessary, we can assume

(145)
$$\int_{\Omega_1} |\nabla u||\nabla \hat{\psi}_i| d\mu < \frac{1}{i}, \quad \int_{\Omega_1} |\nabla u||\nabla \hat{\psi}_i|^q d\mu < \frac{1}{i}$$
$$\text{and} \quad \sup_{\Omega_1} |\nabla \varphi|^{\delta} \int_{\Omega_1} |\nabla \nabla u||\nabla \hat{\psi}_i| d\mu < \frac{1}{i}.$$

Let

$$\psi_i := \varphi^2 \phi^2 \hat{\psi}_i^2.$$

Now notice that since the support of $\nabla \varphi_{J_0}$ is in $B_{2r_{J_0}}(x_{J_0})$ and $|\prod_{J \neq J_0} \varphi_J| \leq 1$, we have

$$\int_{\Omega} |\nabla u||\nabla \varphi| \, d\mu \leq C \int_{\Omega} \left| \sum_{J_0} \nabla \varphi_{J_0} \prod_{J \neq J_0} \varphi_J \right| d\mu$$
$$\leq C \sum_{J_0} \int_{B_{2r_{J_0}}(x_{J_0})} |\nabla \varphi_{J_0}|$$
$$\leq C \sum_{J_0} r_{J_0}^{n-1} < \frac{1}{i}$$

by (142). Similar estimate applies to the integral involving ϕ. Combined with (145), we thus conclude

$$\int_{\Omega} |\nabla u||\nabla \psi_i| \, d\mu$$
$$\leq \int_{\Omega} |\nabla u||\nabla \varphi| \, d\mu + \int_{\Omega} |\nabla u||\nabla \phi| \, d\mu + \int_{\Omega} |\nabla u||\nabla \hat{\psi}_i| \, d\mu$$
(146)
$$\leq \frac{C}{i}$$

which proves inequality (137) of STATEMENT 2[j]. Similar computation proves the inequality

(147)
$$\int_{\Omega} |\nabla u||\nabla \psi_i|^q \, d\mu \leq \frac{C}{i}.$$

We now consider

$$\int_\Omega |\nabla\nabla u||\nabla\psi_i| d\mu = 2\int_\Omega \varphi^2\phi^2\hat\psi_i|\nabla\nabla u||\nabla\hat\psi_i|d\mu + 2\int_\Omega \phi\varphi^2\hat\psi_i^2|\nabla\nabla u||\nabla\phi|d\mu$$
$$+ 2\int_\Omega \varphi\phi^2\hat\psi_i^2|\nabla\nabla u||\nabla\varphi|d\mu$$
$$=: (A) + (B) + (C).$$

Note that by (145),

$$(A) := 2\int_\Omega \varphi^2\phi^2\hat\psi_i|\nabla\nabla u||\nabla\hat\psi_i|d\mu \leq 2\int_{\Omega_1} |\nabla\nabla u||\nabla\hat\psi_i|d\mu < \frac{C}{i}.$$

We next estimate (C). We first note a couple of facts that we will need. First by the order gap of v (cf. Proposition 63),

(148)
$$\sup_{B_{3r_J}(x_J)} |\nabla v| \leq C r_J^{\epsilon_0}.$$

Next, combining the Eells-Sampson and Schoen-Yau formulae (cf. proof of [**GS**] Theorem 6.4), we have

(149)
$$|\nabla\nabla u|^2 |\nabla u|^{-1} \leq C(\triangle|\nabla u| + |\nabla u|) \quad \text{on } \mathcal{R}(u)$$

Now,

$$(C) := 2\int_\Omega \varphi\phi^2\hat\psi_i^2|\nabla\nabla u||\nabla\varphi| \, d\mu$$
$$\leq 2\left(\int_{\cup_{J=1}^l B_{3r_J}(x_J)} \varphi\rho\phi\hat\psi_i|\nabla\varphi|^\delta|\nabla\nabla u|^2|\nabla u|^{-1} \, d\mu\right)^{1/2}$$
$$\times \left(\int_{\cup_{J=1}^l B_{2r_J}(x_J)} |\nabla u||\nabla\varphi|^{2-\delta} \, d\mu\right)^{1/2}$$
$$\leq 2\left(\int_{\cup_{J=1}^l B_{3r_J}(x_J)} \varphi\rho\phi\hat\psi_i|\nabla\varphi|^\delta|\nabla\nabla u|^2|\nabla u|^{-1} \, d\mu\right)^{1/2} \left(C\sum_{J=1}^l r_J^{n-2+\delta}\right)^{1/2}$$
$$\leq 2C \left(\int_{\cup_{J=1}^l B_{3r_J}(x_J)} \varphi\rho\phi\hat\psi_i|\nabla\varphi|^\delta|\nabla\nabla u|^2|\nabla u|^{-1} \, d\mu\right)^{1/2}$$

where the last inequality uses (140) and (142). Noting that the support of the function $\varphi\rho\phi\hat\psi_i|\nabla\varphi|^\delta$ is contained in $\mathcal{R}(u)$, we multiply $\varphi\rho\phi\hat\psi_i|\nabla\varphi|^\delta$ to (149) to obtain

$$\int_{\cup_{J=1}^l B_{3r_J}(x_J)} \varphi\rho\phi\hat\psi_i|\nabla\varphi|^\delta|\nabla\nabla u|^2|\nabla u|^{-1} \, d\mu$$
$$\leq C\int_{\cup_{J=1}^l B_{3r_J}(x_J)} \varphi\rho\phi\hat\psi_i|\nabla\varphi|^\delta|\nabla u| d\mu$$
$$+ C\int_{\cup_{J=1}^l B_{3r_J}(x_J)} \triangle(\varphi\rho\phi\hat\psi_i|\nabla\varphi|^\delta)|\nabla u|d\mu$$
$$=: (C_1) + (C_2).$$

By (140) and (142),

$$(C_1) \leq C \int_{\cup_{J=1}^l B_{3r_J}(x_J)} |\nabla \varphi|^\delta d\mu \leq C \sum_J^l r_J^{n-\delta} \leq \frac{C}{i}.$$

To estimate (C_2), we claim

(150) $$||\nabla u| - |\nabla V|| \leq C|\nabla v|.$$

To justify (150), we use the mean value theorem to write

$$\frac{(|\nabla V|^2 + s)^{\frac{1}{2}} - |\nabla V|}{s} = \frac{1}{2}(|\nabla V|^2 + c)^{-\frac{1}{2}}$$

for some $c \in (0, s)$. Letting $s = |\nabla v|^2 + 2 <\nabla V, \nabla v>$, we have

$$|\nabla u| = |\nabla V| + \frac{1}{2}(|\nabla V|^2 + c)^{-\frac{1}{2}}(|\nabla v|^2 + 2 <\nabla V, \nabla v>).$$

In the support of ρ and φ which is contained in $\mathcal{N}$, (141) implies

$$(|\nabla V|^2 + c)^{-\frac{1}{2}} \leq |\nabla V|^{-1} \leq C.$$

which then implies (150). By (150), we have that

$$\int_\Omega \varphi\rho\phi|\nabla\varphi|^\delta \Delta \hat{\psi}_i (|\nabla u| - |\nabla V|) \, d\mu$$
$$= -\int_\Omega (|\nabla u| - |\nabla V|) \nabla(\varphi\rho\phi|\nabla\varphi|^\delta) \cdot \nabla \hat{\psi}_i d\mu$$
$$-\int_\Omega \varphi\rho\phi|\nabla\varphi|^\delta \nabla \hat{\psi}_i \cdot \nabla(|\nabla u| - |\nabla V|) \, d\mu$$
$$\leq C \left(\int_\Omega |\nabla(\varphi\rho\phi|\nabla\varphi|^\delta)|^p |\nabla v| d\mu\right)^{\frac{1}{p}} \left(\int_\Omega |\nabla \hat{\psi}_i|^q |\nabla v| d\mu\right)^{\frac{1}{q}}$$
$$+C \int_\Omega |\nabla\varphi|^\delta |\nabla \hat{\psi}_i||\nabla\nabla u| d\mu.$$

Therefore,

$$
\begin{aligned}
(C_2) &:= \int_{\cup_{J=1}^{l} B_{3r_J}(x_J)} \triangle(\varphi \rho \phi \hat{\psi}_i |\nabla \varphi|^\delta) |\nabla u| d\mu \\
&= \int_{\cup_{J=1}^{l} B_{3r_J}(x_J)} \triangle(\varphi \rho \phi \hat{\psi}_i |\nabla \varphi|^\delta) \left(|\nabla V| + |\nabla u| - |\nabla V| \right) d\mu \\
&= \int_\Omega \triangle(\varphi \rho \phi \hat{\psi}_i |\nabla \varphi|^\delta) |\nabla V| d\mu + \int_\Omega \hat{\psi}_i \triangle(\varphi \rho \phi |\nabla \varphi|^\delta) \left(|\nabla u| - |\nabla V| \right) d\mu \\
&\quad + \int_\Omega (\varphi \rho \phi |\nabla \varphi|^\delta) \triangle \hat{\psi}_i \left(|\nabla u| - |\nabla V| \right) d\mu \\
&\quad + 2 \int_\Omega \nabla(\varphi \rho \phi |\nabla \varphi|^\delta) \cdot \nabla \hat{\psi}_i \left(|\nabla u| - |\nabla V| \right) d\mu \\
&\leq C \left(\int_\Omega |\nabla(\varphi \rho \phi \hat{\psi}_i |\nabla \varphi|^\delta)|^q d\mu \right)^{\frac{1}{q}} \cdot \left(\int_\Omega |\nabla \nabla V|^p d\mu \right)^{\frac{1}{p}} \\
&\quad + C \int_\Omega \hat{\psi}_i \left| \triangle(\varphi \rho \phi |\nabla \varphi|^\delta) \right| |\nabla v| d\mu \\
&\quad + C \left(\int_\Omega |\nabla(\varphi \rho \phi |\nabla \varphi|^\delta)|^p |\nabla v| d\mu \right)^{\frac{1}{p}} \left(\int_\Omega |\nabla \hat{\psi}_i|^q |\nabla v| d\mu \right)^{\frac{1}{q}} \\
&\quad + C \int_\Omega \varphi \rho \phi |\nabla \varphi|^\delta |\nabla \hat{\psi}_i| |\nabla \nabla u| d\mu \\
&=: (C_{21}) + (C_{22}) + (C_{23}) + (C_{24}).
\end{aligned}
$$

Using the fact that $|\nabla \nabla V| \in L^p(\Omega)$ (cf. Lemma 50) and the fact that derivatives of φ and ρ are supported in $\cup_{J=1}^{l} B_{3r_J}(x_J)$ and the derivatives of ϕ are supported in $\cup_{J=1}^{l'} B_{3s_J}(\xi_J)$, we have

$$
\begin{aligned}
(C_{21}) &\leq C \left(\int_\Omega |\nabla(\varphi \rho \phi |\nabla \varphi|^\delta)|^q d\mu \right)^{\frac{1}{q}} \\
&\leq C \left(\sum_{J=1}^{l} r_J^{n-q-q\delta} + \sup_\Omega |\nabla \varphi|^{\delta q} \sum_{J=1}^{l'} s_J^{n-q} \right)^{\frac{1}{q}} \\
&\leq C \left(\frac{1}{i} \right)^{\frac{1}{q}}
\end{aligned}
$$

by (140), (142) and (144). Similar argument along (145) gives

$$
\begin{aligned}
(C_{23}) &\leq C \left(\int_\Omega |\nabla(\varphi \rho \phi |\nabla \varphi|^\delta)|^p |\nabla v| d\mu \right)^{\frac{1}{p}} \\
&\leq C \left(\sum_{J=1}^{l} r_J^{n-p-p\delta+\epsilon_0} + \sup_\Omega |\nabla \varphi|^{\delta p} \sum_{J=1}^{l'} s_J^{n-p+\epsilon_0} \right)^{\frac{1}{p}} \\
&< C \left(\frac{1}{i} \right)^{\frac{1}{p}}
\end{aligned}
$$

and the same way
$$(C_{22}) < C\left(\frac{1}{i}\right)^{\frac{1}{p}}.$$

Finally, (145) also yields
$$(C_{24}) \leq C\sup_{\Omega}|\nabla\varphi|^{\delta}\int_{\Omega}|\nabla\hat{\psi}_i||\nabla\nabla u|d\mu < \frac{C}{i}.$$

Combining the estimates for (C_{21}), (C_{22}), (C_{23}), (C_{24}) with that of (C_1), we obtain $(C) \leq \frac{C}{i}$. The estimate for (B) is analogous to (C) but simpler. We repeat the argument in (C) with $\delta = 0$ keeping in mind that, since we are near higher order points of u, we have

(151)
$$\sup_{B_{2s_J}(\xi_J)}|\nabla u| \leq Cs_J^{\epsilon_0}$$

the order gap of u of Proposition 62 (along with the monotonicity property of u, cf. proof of [**GS**] Theorem 2.4). (Note that in (C) we only have $|\nabla u|$ bounded near $\mathcal{S}_j(u)$.) Combining the estimates for (A), (B) and (C), we obtain

(152)
$$\int_{\Omega}|\nabla\nabla u||\nabla\psi_i|d\mu \leq C\left(\frac{1}{i}\right)^{\frac{1}{p}}.$$

The inequalities (146), (147) and (152) show that STATEMENT $2[j]$ holds. □

The above completes the proof of Theorem 1 and Theorem 2. The inductive process also yields Theorem 3 as a consequence of Proposition 54. Similarly, Theorem 4 is an immediately consequence of Proposition 63. Furthermore, from Corollary 60, we can immediately deduce the following:

COROLLARY 64. *If* $u = (V, v) : B_{\sigma_\star}(x_\star) \to (\mathbf{R}^j \times Y_2^{k-j}, d_G)$ *is a harmonic map, then there exist* $C > 0$, $c > 0$, $R_0 > 0$ *and* $\epsilon_0 > 0$ *such that*

$$1 + \epsilon_0 \leq e^{c\omega^2}\frac{\sigma E_{x_0}^v(\sigma)}{I_{x_0}^v(\sigma)} \leq C, \quad \frac{I_{x_0}^v(\sigma)}{\sigma^{n+1+2\epsilon_0}} \leq C, \quad \frac{E_{x_0}^v(\sigma)}{\sigma^{n+2+2\epsilon_0}} \leq C$$

for all $x_0 \in \mathcal{S}_j(u) \cap B_{\frac{\sigma_\star}{2}}(x_\star)$ *and* $\sigma \in (0, R_0)$.

APPENDIX A

Appendix 1

The goal of this appendix is to establish Proposition 67 below which is an analogue of [GS] Theorem 5.1. Recall that in Section 11, Proposition 67 was applied to the singular component map v of a harmonic map into a DM-complex and $x_0 \in \mathcal{S}_j(u)$. The main difference from [GS] is that the map v is not necessarily harmonic but only approximately harmonic.

We start with a preliminary lemma.

LEMMA 65. *Let* $u = (V,v) : B_{\sigma_\star}(x_\star) \to (\mathbf{R}^j \times Y_2^{k-j}, d_G)$ *be a harmonic map as in (16). Let* $\rho, \vartheta \in (0,1]$. *For* $x_0 \in B_{\frac{\sigma_\star}{2}}(x_\star) \cap \mathcal{S}_j(u)$ *and* $\{v_\sigma\}$ *the blow up maps of* v *at* x_0, *there exists a constant* $C > 0$ *depending only on* ρ, *the constant in the estimates (28) and (29) for the target metric* G, *the domain metric* g *and the Lipschitz constant of* u *such that for any harmonic map*

$$w : (B_{\vartheta\rho}(0), g_\sigma) \to Y_2^{k-j} \text{ with } E^w(\vartheta\rho) \leq E^{v_\sigma}(\vartheta\rho),$$

we have

(153) $$\sup_{B_{\frac{\rho\vartheta}{4}}(0)} d^2(v_\sigma(x), w(x)) \leq C \int_{\partial B_{\rho\vartheta}(0)} d^2(v_\sigma, w) d\Sigma_\sigma + C\sigma^2 \vartheta^{4i}.$$

PROOF. Let $\hat{w}(x) : B_{\sigma\vartheta}(0) \to Y_2^{k-j}$ be $\hat{w}(x) = \nu_\sigma \tilde{w}(\sigma^{-1}x)$. Note that as in Proposition 47 the Lipschitz constants of $\hat{w}$ are uniformly bounded. Rewriting (80), we have

$$-C \int_{B_{\sigma\vartheta}(0)} \eta d(v, P_0) d(v, \hat{w}) d\mu \leq -\int_{B_{\sigma\vartheta}(0)} \nabla \eta \cdot \nabla d^2(v, \hat{w}) \, d\mu.$$

Let $x \in B_{\frac{\sigma\rho\vartheta}{4}}(0)$ and η approximate the characteristic function of $B_s(x) \subset B_{\sigma\rho\vartheta}(0)$ to obtain

$$-C \int_{B_s(x)} d(v, P_0) d(v, \hat{w}) d\mu \leq \int_{\partial B_s(x)} \frac{\partial}{\partial s} d^2(v, \hat{w}) \, d\mu.$$

As in Proposition 48

$$\int_{\partial B_s(x)} \frac{\partial}{\partial s} d^2(v, \hat{w}) d\Sigma \leq s^{n-1} \frac{d}{ds} \left(\frac{e^{Cs^2}}{s^{n-1}} \int_{\partial B_s(x)} d^2(v, \hat{w}) d\Sigma \right).$$

Combining the above two inequalities for $s \in (0, \frac{\sigma\rho\vartheta}{4})$,

(154) $$0 \leq \frac{d}{ds} \left(\frac{e^{Cs^2}}{s^{n-1}} \int_{\partial B_s(x)} d^2(v, \hat{w}) d\Sigma \right) + Cs^{-n+1} \int_{B_{\frac{\sigma\rho\vartheta}{4}}(x)} d(v, \hat{w}) d(v, P_0) d\mu.$$

A. APPENDIX 1

Integrating this over $s \in (0, t)$, we obtain

$$0 \leq \frac{e^{Ct^2}}{t^{n-1}} \int_{\partial B_t(x)} d^2(v, \hat{w}) d\Sigma - \frac{1}{C_n} d^2(v(x), \hat{w}(x)) + Ct^{-n+2} \int_{B_{\frac{\sigma\rho\vartheta}{4}}(x)} d(v, \hat{w}) d(v, P_0) d\mu.$$

Thus,

$$t^{n-1} d^2(v(x), \hat{w}(x)) \leq C \int_{\partial B_t(x)} d^2(v, \hat{w}) d\Sigma + Ct \int_{B_{\frac{\sigma\rho\vartheta}{4}}(x)} d^2(v, \hat{w}) d(v, P_0) d\mu.$$

Integrating this over $t \in (0, \frac{\sigma\rho\vartheta}{4})$, we obtain

$$d^2(v(x), \hat{w}(x)) \leq \frac{C}{(\sigma\vartheta)^n} \int_{B_{\frac{\sigma\rho\vartheta}{4}}(x)} d^2(v, \hat{w}) d\mu + \frac{C}{(\sigma\vartheta)^{n-2}} \int_{B_{\frac{\sigma\rho\vartheta}{4}}(x)} d(v, \hat{w}) d(v, P_0) d\mu.$$

Since

$$x \in B_{\frac{\sigma\rho\vartheta}{4}}(0) \Rightarrow B_{\frac{\sigma\rho\vartheta}{4}}(x) \subset B_{\sigma\rho\vartheta}(0),$$

we obtain

$$\sup_{B_{\frac{\sigma\rho\vartheta}{4}}(0)} d^2(v(x), \hat{w}(x))$$

$$\leq \frac{C}{(\sigma\vartheta)^n} \int_{B_{\sigma\rho\vartheta}(0)} d^2(v, \hat{w}) d\mu + \frac{C}{(\sigma\vartheta)^{n-2}} \int_{B_{\sigma\rho\vartheta}(0)} d(v, \hat{w}) d(v, P_0) d\mu$$

(155) $$\leq \frac{C}{(\sigma\vartheta)^n} \int_{B_{\sigma\rho\vartheta}(0)} d^2(v, \hat{w}) d\mu + \frac{C}{(\sigma\vartheta)^{n-2}} \int_{B_{\sigma\rho\vartheta}(0)} d^2(v, P_0) d\mu$$

where we have used the Cauchy-Schwartz inequality and adjusted the constant C in the last inequality. Similarly to (154), we have for $s \in (0, \rho\sigma\vartheta)$

$$0 \leq \frac{d}{ds} \left(\frac{e^{Cs^2}}{s^{n-1}} \int_{\partial B_s(0)} d^2(v, \hat{w}) d\Sigma \right) + Cs^{-n+1} \int_{B_{\rho\sigma\vartheta}(0)} d(v, \hat{w}) d(v, P_0) d\mu.$$

Integrating this over $s \in (t, \sigma\rho\vartheta)$, we obtain

$$0 \leq \frac{e^{C\rho^2\theta^{2i}}}{(\sigma\rho\vartheta)^{n-1}} \int_{\partial B_{\sigma\rho\vartheta}(0)} d^2(v, \hat{w}) d\Sigma - \frac{e^{Ct^2}}{t^{n-1}} \int_{\partial B_t(0)} d^2(v, \hat{w}) d\Sigma$$

$$+ C(\sigma\rho\vartheta)^{(-n+2)} \int_{B_{\sigma\rho\vartheta}(0)} d(v, \hat{w}) d(v, P_0) d\mu,$$

and hence

$$\int_{\partial B_t(0)} d^2(v, \hat{w}) d\Sigma$$

$$\leq \frac{Ct^{n-1}}{(\rho\sigma\vartheta)^{n-1}} \int_{\partial B_{\rho\sigma\vartheta}(0)} d^2(v, \hat{w}) d\Sigma + \frac{Ct^{n-1}}{(\rho\sigma\vartheta)^{n-2}} \int_{B_{\rho\sigma\vartheta}(0)} d(v, \hat{w}) d(v, P_0) d\mu.$$

Furthermore, integrating this over $t \in (0, \rho\sigma\vartheta)$, we obtain

$$\frac{1}{(\sigma\rho\vartheta)^n} \int_{B_{\sigma\rho\vartheta}(0)} d^2(v, \hat{w}) d\mu$$

$$\leq \frac{C}{(\sigma\rho\vartheta)^{n-1}} \int_{\partial B_{\sigma\rho\vartheta}(0)} d^2(v, \hat{w}) d\Sigma + \frac{C}{(\sigma\rho\vartheta)^{n-2}} \int_{B_{\sigma\rho\vartheta}(0)} d(v, \hat{w}) d(v, P_0) d\mu$$

which implies (by Cauchy-Schwartz and adjusting the constant C)

(156) $$\frac{1}{(\sigma\vartheta)^n}\int_{B_{\sigma\rho\vartheta}(0)}d^2(v,\hat{w})d\mu$$
$$\leq \frac{C}{(\sigma\vartheta)^{n-1}}\int_{\partial B_{\sigma\rho\vartheta}(0)}d^2(v,\hat{w})d\Sigma + \frac{C}{(\sigma\vartheta)^{n-2}}\int_{B_{\sigma\rho\vartheta}(0)}d^2(v,P_0)d\mu.$$

Combining (155) and (156), we obtain
$$\sup_{B_{\frac{\rho\vartheta}{4}}(0)} d^2(v(x),\hat{w}(x))$$
$$\leq \frac{C}{(\sigma\vartheta)^{n-1}}\int_{\partial B_{\sigma\rho\vartheta}(0)}d^2(v,\hat{w})d\Sigma + \frac{C}{(\sigma\vartheta)^{n-2}}\int_{B_{\sigma\rho\vartheta}(0)}d^2(v,P_0)d\mu.$$

Multiplying by ν_σ^{-1} and applying change of variables, we obtain

$$\sup_{B_{\frac{\rho\vartheta}{4}}(0)} d^2(v_\sigma(x),w(x)) \leq C\int_{\partial B_\rho(0)} d^2(v_\sigma,w)d\Sigma_\sigma$$

(157) $$+C\sigma^2\frac{C}{\vartheta^{n-2}}\int_{B_{\rho\vartheta}(0)} d^2(v_\sigma,P_0)d\mu_\sigma.$$

The monotonity formula of Corollary 60 implies
$$\frac{\int_{\partial B_s(0)}d^2(v_\sigma,P_0)d\Sigma_\sigma}{s^{n-1}} = \nu_\sigma^{-2}\frac{\int_{\partial B_{\sigma s}(0)}d^2(v,P_0)d\Sigma}{(\sigma s)^{n-1}}$$
$$= \nu_\sigma^{-2}(\sigma s)^2\frac{\int_{\partial B_{\sigma s}(0)}d^2(v,P_0)d\Sigma}{(\sigma s)^{n+1}}$$
$$\leq e^C\nu_\sigma^{-2}(\sigma s)^2\frac{\int_{\partial B_\sigma(0)}d^2(v,P_0)d\Sigma}{\sigma^{n+1}}$$
$$= e^C\nu_\sigma^{-2}s^2\frac{\int_{\partial B_\sigma(0)}d^2(v,P_0)d\Sigma}{\sigma^{n-1}}$$
$$= e^C s^2\int_{\partial B_1(0)}d^2(v_\sigma,P_0)d\Sigma_\sigma$$
$$= e^C s^2.$$

Thus,
$$\int_{\partial B_s(0)}d^2(v_\sigma,P_0)d\Sigma_\sigma \leq Cs^{n+1}.$$

Integrating this with respect to s over the interval $(0,\rho\vartheta)$, we conclude that the second term on the right hand side of (157) is bounded by $C\sigma^2\vartheta^4$ implies the assertion of the lemma. □

REMARK 66. As is the case in Chapter 6, the properties of u that we need in the proof of Lemma 65 are Assumption 2, Assumption 3 and Assumption 4 of Chapter 5.

PROPOSITION 67. *Let metrics G and h defined on $\mathbf{R}^j \times Y_2^{k-j}$ and Y_2^{k-j} and $u = (V,v) : B_{\sigma_*}(x_*) \to (\mathbf{R}^j \times Y_2^{k-j}, d_G)$ be a harmonic map that satisfy Assumption 2, Assumption 3 and Assumption 4 of Chapter 5. Let $x_0 \in B_{\frac{\sigma_*}{2}}(x_*) \cap \mathcal{S}_j(u)$, $\{v_\sigma\}$ the blow up maps of v at x_0 and $l : B_1(0) \to Y_2^{k-j}$ a homogeneous degree 1 map. There exists $D_0 > 0$ such that*

$$\sup_{B_{\frac{7}{8}}(0)} d(v_\sigma, l) < D_0,$$

then there exists $\lambda > 0$ such that

$$\sup_{B_t(0)} d(v_\sigma, P_0) > \lambda t$$

for t sufficiently small.

As before, we let $g_\theta(x) = g(\theta x)$ and $d\Sigma_\theta$, $d\mu_\theta$ be the volume forms on $\partial B_\rho(0)$, $B_\rho(0)$ respectively. For notational simplicity, we will sometimes omit the subscript and write $d\Sigma$ and $d\mu$.

The proof of Proposition 67 involves a proof by induction described in the following lemma.

LEMMA 68. *Let metrics G and h defined on $\mathbf{R}^j \times Y_2^{k-j}$ and Y_2^{k-j} and $u = (V,v) : B_{\sigma_*}(x_*) \to (\mathbf{R}^j \times Y_2^{k-j}, d_G)$ be a harmonic map that satisfy Assumption 2, Assumption 3 and Assumption 4 of Chapter 5 and $l : B_1(0) \to Y_2^{k-j}$ a homogeneous degree map. For $\vartheta \in (0,1]$, let $v^\vartheta(x) = \vartheta^{-1} v(\vartheta x)$ and $l^\vartheta(x) = \theta^{-1} l(\vartheta x)$. Assume the following conditions:*

(i) *For $\rho \in [\frac{3}{4}, \frac{7}{8}]$ and the harmonic map*

$$w : (B_\rho(0), g_\vartheta) \to Y_2^{k-j} \text{ with } w|_{\partial B_\rho(0)} = v^\vartheta|_{\partial B_\rho(0)}$$

we have

(158) $$\sup_{B_{\frac{1}{4}}(0)} d^2(v^\vartheta(x), w(x)) \le C \int_{\partial B_\rho(0)} d^2(v^\vartheta, w) d\Sigma + c\vartheta^2.$$

(ii) *For $w : B_\rho(0) \subset \mathbf{R}^n \to Y_2^{k-j}$ as in (i) there exists $\beta > 0$ and a homogeneous degree 1 map $\hat{l} : B_1(0) \subset \mathbf{R}^n \to Y_2^{k-j}$ such that*

(159) $$\sup_{B_\theta(0)} d(w, \hat{l}) \le C\theta^{1+\beta} \inf_L \sup_{B_{\frac{1}{2}}(0)} d(w, L), \forall \theta \in (0, \frac{1}{4})$$

where the infimum is taken over all homogeneous maps L of degree 1.

(iii) *The constants $C > 1$, $\theta \in (0, \frac{1}{4})$, β, $c \in (0,1)$ and the map l^1 satisfy*

(160) $$C\theta^\beta < \frac{1}{8},$$

and

(161) $$c\theta^{-2} < \frac{D_0^2}{32}.$$

A. APPENDIX 1

Let $_il : B_1(0) \subset \mathbf{R}^n \to Y_2^{k-j}$ be a homogeneous map of degree 1 then we have the following implication:

(162) $$\begin{cases} \sup_{B_1(0)} d(v^{\theta^i}, {}_il) < \dfrac{D_0}{2^i} \\ \sup_{B_1(0)} d(v^{\theta^i}, l^{\theta^i}) d\mu < {}_i\delta \end{cases}$$

implies that there exists a homogeneous degree 1 map $_{i+1}l : B_1(0) \to Y_2^{k-j}$ so that

(163) $$\begin{cases} \sup_{B_1(0)} d(v^{\theta^{i+1}}, {}_{i+1}l) < \dfrac{D_0}{2^{i+1}} \\ \sup_{B_1(0)} d(v^{\theta^{i+1}}, l^{\theta^{i+1}}) < {}_{i+1}\delta := 2\theta^{-1}\dfrac{D_0}{2^i} + {}_i\delta. \end{cases}$$

PROOF. We first give the proof of the first inequality of (163). Let $w : B_\rho(0) \to Y_2^{k-j}$ be given as in (i) with $\vartheta = \theta^i$. First, since $w|_{\partial B_\rho(0)} = v^{\theta^i}|_{\partial B_\rho(0)}$,

$$\sup_{B_{\frac{1}{4}}(0)} d^2(v^{\theta^i}, w) \leq c\theta^{2i} \quad \text{(by (158))}$$

(164) $$\leq \dfrac{\theta^2}{16}\left(\dfrac{D_0}{2^i}\right)^2 \quad \text{(by (161))}.$$

By Assumption (ii) inequality (159), there exists a homogeneous degree 1 harmonic map $\hat{l} : B_1(0) \to Y_2^{k-j}$ such that

(165) $$\sup_{B_\theta(0)} d(w, \hat{l}) \leq C\theta^{1+\beta} \sup_{B_{\frac{1}{2}}(0)} d(w, {}_il).$$

With

$$_{i+1}l : B_1(0) \to Y_2^{k-j} \text{ defined by } _{i+1}l(x) = {}^\theta\hat{l}(x),$$

we obtain

$$\sup_{B_1(0)} d(w^\theta, {}_{i+1}l) \leq \theta^{-1} \sup_{B_{\frac{1}{4}}(0)} d(w, \hat{l})$$

$$\leq C\theta^\beta \sup_{B_{\frac{1}{4}}(0)} d(w, {}_il) \quad \text{(by (165))}$$

$$\leq C\theta^\beta \left(\sup_{B_{\frac{1}{4}}(0)} d(w, v^{\theta^i}) + \sup_{B_{\frac{1}{4}}(0)} d(v^{\theta^i}, {}_il)\right)$$

$$< C\theta^\beta \left(\dfrac{\theta}{4}\dfrac{D_0}{2^i} + \dfrac{D_0}{2^i}\right) \quad \text{(by (164) and (162))}$$

$$< C\theta^\beta \dfrac{D_0}{2^{i-1}}$$

$$< \dfrac{1}{2}\dfrac{D_0}{2^{i+1}} \quad \text{(by (160))}.$$

Combined with (164), this implies

$$\sup_{B_1(0)} d(v^{\theta^{i+1}}, {}_{i+1}l) \leq \sup_{B_1(0)} d(v^{\theta^{i+1}}, w^\theta) + \sup_{B_1(0)} d(w^\theta, {}_{i+1}l) < \dfrac{D_0}{2^{i+1}}.$$

This completes the proof of the first inequality of (163).

We now prove the second inequality of (163). Since $l^{\theta^i}(0) = v^{\theta^i}(0)$, we have

$$d(l^{\theta^i}(0), {}_il(0)) = d(v^{\theta^i}(0), {}_il(0)) < \frac{D_0}{2^i}.$$

By the NPC condition, we obtain for any $x \in B_1(x)$ that

$$d(l^{\theta^i}(\theta x), {}_il(\theta x)) \leq (1-\theta)d(l^{\theta^i}(0), {}_il(0)) + \theta d(l^{\theta^i}(x), {}_il(x)).$$

Thus,

$$\begin{aligned}
d(l^{\theta^i}(\theta x), {}_il(\theta x)) &< (1-\theta)\frac{D_0}{2^i} + \theta d(l^{\theta^i}(x), {}_il(x)) \\
&\leq (1-\theta)\frac{D_0}{2^i} + \theta(d(l^{\theta^i}(x), v^{\theta^i}(x)) + d(v^{\theta^i}(x), {}_il(x))) \\
&< (1-\theta)\frac{D_0}{2^i} + \theta({}_i\delta + \frac{D_0}{2^i}) \\
&\leq \frac{D_0}{2^i} + \theta \, {}_i\delta.
\end{aligned}$$

Therefore,

$$d(v^{\theta^{i+1}}(x), l^{\theta^{i+1}}(x)) < 2\theta^{-1}\frac{D_0}{2^i} + {}_i\delta.$$

This proves the second inequality of (163) and completes the proof. $\square$

PROOF OF PROPOSITION 67. Assume

$$\sup_{B_{\frac{1}{2}}(0)} d(v_\sigma, l) < D_0.$$

For x_1 is sufficiently close to 0 such that setting $Q = v(x_1)$ and choosing new coordinates with x_1 identified as 0 via normal coordinates centered at x_1, we have

$$\sup_{B_{\frac{1}{4}}(0)} d(v_\sigma, l_Q) < D_0.$$

Define $v^1 : B_1(0) \to Y_2^{k-j}$ and $l^1 : B_1(0) \to Y_2^{k-j}$ by setting $v^1(x) = v_\sigma(\frac{x}{4})$ and $l^1(x) = l_Q(\frac{x}{4})$ and hence

$$\sup_{B_1(0)} d(v^1, l^1) < D_0.$$

Choose $C > 0$ sufficiently large such that inequality (158) of Assumption (i) and the inequality (159) of Assumption (ii) are both valid. This can be done by Lemma 65 and [**GS**] Theorem 6.3. For this choice of C, to obtain inequality (158) choose $\theta \in (0, \frac{1}{4})$ such that (160) is satisfied. Finally, choose σ_0 sufficiently small such that (161) is satisfied for $c = C\sigma_0^2$. With this choice of c, we can rewrite the inequality (158) as

$$\sup_{x \in B_{\frac{1}{4}}(0)} d^2(v_\sigma^{\theta^i}(x), w(x)) \leq C \int_{\partial B_\rho(0)} d^2(v_\sigma^{\theta^i}, w) d\Sigma + c\theta^{2i}$$

for a harmonic map $w : (B_\rho(0), g_{\theta^i \sigma}) \to Y_2^{k-j}$ with $\rho \in [\frac{3}{4}, \frac{7}{8}]$. In other words, condition (i) of Lemma 68 is satisfied for v_σ. By [**GS**] Theorem 6.3, condition (ii) of Lemma 68 is satisfied. With the choices of constants above, condition (iii) of Lemma 68 is satisfied.

Apply Lemma 68 to obtain

$$\sup_{B_1(0)} d(v^{\theta^i}, l^{\theta^i}) <\ _i\delta$$

$$= \theta^{-1}\frac{D_0}{2^{i-1}} +\ _{i-1}\delta$$

$$= \theta^{-1}\sum_{j=0}^{i-1}\frac{D_0}{2^j} +\ _0\delta$$

$$\leq 2\theta^{-1}D_0 + D_0.$$

Thus,
$$\sup_{B_{\theta^i}(x)} d(v_\sigma, l) < 2\theta^{i-1}D_0 + \theta^i D_0.$$

For $s > 0$, let j such that $\theta^{j+1} \leq s < \theta^j$. Then

$$\sup_{B_s(0)} d(v_\sigma, l) < 2\theta^{j-1}D_0 + \theta^j D_0 < (2\theta^{-2}D_0 + \theta^{-1}D_0)s.$$

Assume D_0 is sufficiently small such that

$$\sup_{B_s(0)} d(v_\sigma, P_0) \geq \sup_{B_s(0)} d(l, P_0) - \sup_{B_s(0)} d(v_\sigma, l) > \lambda s$$

for some $\lambda > 0$. □

APPENDIX B

Appendix 2

The purpose of this Appendix is provide a proof of the crucial codimension 2 property for a set of higher order points needed in the proof of Theorem 1. As described in the proof of Theorem 1, we need two separate statements: one for the original harmonic map u and one for the singular component v. In addition, a more general statement is needed in future applications. Thus, we will prove a general codimension 2 statement that covers all cases at once. We start with lemma regarding the upper semicontinuity of Hausdorff dimension.

LEMMA 69. *If S_i be a sequence of closed subsets of $B_1(0)$ satisfying a property that*

(166) $$x_i \in S_i \text{ and } x_i \to x_0 \in B_1(0) \Rightarrow x_0 \in S_0$$

for some closed subset S_0 of $B_1(0)$, then

(167) $$\limsup_{i \to \infty} \dim_{\mathcal{H}}(S_i) \leq \dim_{\mathcal{H}}(S_0).$$

PROOF. Following [**GS**], define $\hat{\mathcal{H}}^s(\cdot)$ by

$$\hat{\mathcal{H}}^s(S) = \inf \left\{ \sum_{l=1}^{\infty} r_l^s : \text{ all coverings } \{B_{r_l}(x_l)\}_{l=1}^{\infty} \text{ of } S \text{ by open balls} \right\}.$$

Called the rough outer Hausdorff measure, $\hat{\mathcal{H}}^s$ is not precisely the Hausdorff measure $\mathcal{H}^s$, but its importance is in the fact that the Hausdorff dimension of any set S is given by

$$\dim_{\mathcal{H}}(S) = \inf\{s : \mathcal{H}^s(S) = 0\} = \inf\{s : \hat{\mathcal{H}}^s(S) = 0\}.$$

We now come to the proof of (167). First, fix $s > 0$ and let $r \in (0,1)$. Given $\epsilon_1 > 0$, let $\{B_{r_l}(x_l)\}_{l=1}^N$ be a finite covering of $S_0 \cap \overline{B_r(0)}$ such that $x_l \in S_0$ and

$$\hat{\mathcal{H}}^s(S_0 \cap \overline{B_r(0)}) + \epsilon_1 \geq \sum_{l=1}^N r_l^s.$$

Note here that it is enough to consider <u>finite</u> coverings since S_0 is compact. By (166), $\{B_{r_l}(x_l)\}_{l=1}^N$ is a covering of $S_i \cap \overline{B_r(0)}$ for i sufficiently large. Hence, for i sufficiently large,

$$\hat{\mathcal{H}}^s(S_0 \cap \overline{B_r(0)}) + \epsilon_1 \geq \sum_{l=1}^N r_l^s \geq \hat{\mathcal{H}}^s(S_i \cap \overline{B_r(0)}).$$

Since ϵ_1 is arbitrary, this proves (167). □

Recall that we are interested in maps that are not necessarily harmonic. More precisely, we are interested in maps given in the following

DEFINITION 70. Let $v : B_{\sigma_*}(x_*) \to (Y,d)$ be a finite energy continuous map from a Riemannian domain into an NPC space and let $\mathcal{S}$ be a closed subset of $B_{\frac{\sigma_*}{2}}(x_*)$. For $x \in \mathcal{S}$ and $0 < \sigma < \sigma_0 =: \sup\{\sigma : B_\sigma(x) \subset B_{\sigma_*}(x_*)\}$, assume that v is not constant in any neighborhood of x and define

$$Ord^v(x,\sigma) := \frac{\sigma E_x^v(\sigma)}{I_x^v(\sigma)}.$$

We say v *satisfies (P1), (P2) and (P3) with respect to $\mathcal{S}$* if it satisfies the properties below.

(P1) At any $x \in \mathcal{S}$, we require that v has a well defined order at x in the sense that it satisfies the following properties: there exist constants $C > 0$, $C_1 > 0$ and $R_0 > 0$ such that for any $x \in \mathcal{S}$, there exists a function $\sigma \mapsto J_x(\sigma)$ satisfying

$$e^{-C_1\sigma} I_x^v(\sigma) \leq J_x(\sigma) \leq I_x^v(\sigma) e^{C_1\sigma}, \ \forall \sigma \in (0, R_0),$$

$$\sigma \mapsto e^{C\sigma} \frac{\sigma E_x^v(\sigma)}{J_x(\sigma)} \text{ is non-decreasing in } (0, R_0),$$

$$\lim_{\sigma \to 0} Ord^v(x) := \lim_{\sigma \to 0} Ord^v(x,\sigma) \text{ exists}$$

and

$$Ord^v(x) \leq e^{(C+C_1)\sigma} \frac{\sigma E_x^v(\sigma)}{I_x^v(\sigma)}, \ \forall \sigma \in (0, R_0).$$

(P2) For any $x \in \mathcal{S}$, identify $x = 0$ via normal coordinates and define *blow-up maps* and *approximating blow-up maps* at x as follows. We first define the restriction maps

$$_\sigma v : (B_\sigma(0), g) \to Y, \ _\sigma v = v\big|_{B_\sigma(0)},$$

the harmonic maps

$$_\sigma w : (B_\sigma(0), g) \to (Y, d) \text{ with } _\sigma w\big|_{\partial B_\sigma(0)} = _\sigma v\big|_{\partial B_\sigma(0)}$$

and set

$$\nu_\sigma = \left(\frac{I_0^{\sigma v}(\sigma)}{\sigma^{n-1}} \right)^{1/2}. \tag{168}$$

Let $g_\sigma(y) = g(\sigma y)$ be the rescaled metric on $B_1(0)$ and define the rescaled maps

$$v_\sigma, w_\sigma : (B_1(0), g_\sigma) \to (Y, \nu_\sigma^{-1} d)$$

by setting

$$v_\sigma(y) = _\sigma v(\sigma y) \text{ and } w_\sigma(y) = _\sigma w(\sigma y).$$

The normalization by ν_σ implies that

$$I_0^{v_\sigma,x}(1) = 1.$$

We require that given a sequence $\sigma_i \to 0$, there exists a subsequence (which we call again σ_i by a slight abuse of notation) such that the blow up maps $v_{\sigma_i} : B_1(0) \to (Y, \nu^{-1}(\sigma_i) d)$ at x converge locally uniformly in the pullback sense to a homogeneous harmonic map $v_0 : (B_1(0), \delta) \to (Y_0, d_0)$ for some NPC space . We also require that for any $r \in (0,1)$

$$\lim_{i \to \infty} \sup_{B_r(0)} d(v_{\sigma_i}, w_{\sigma_i}) = 0.$$

In particular, w_{σ_i} also converges locally uniformly in the pullback sense to v_0. Furthermore, for any $\sigma_i \xi \in \mathcal{S}$ we have $Ord^{v_{\sigma_i}}(\xi) = Ord^v(\sigma_i \xi)$. In other words, the order for v_{σ_i} exists for any point in $\sigma_i^{-1}\mathcal{S}$.

(P3) With the notation as in (P2), we require that for the sequences v_{σ_i}, w_{σ_i} in (P2) and for any $R \in (0,1)$, there exists a constant $C > 0$ such that for any $\xi \in B_R(0)$ and $r > 0$ such that $B_r(\xi) \subset B_R(0)$,

$$\left| E^{v_{\sigma_i}}_\xi(r) - E^{w_{\sigma_i}}_\xi(r) \right| < C\sigma_i. \tag{169}$$

REMARK 71. A harmonic map $u : B_1(0) \to Y$ into an NPC space satisfies properties (P1), (P2) and (P3) with respect to $\mathcal{S} = B_{\frac{\sigma_*}{2}}(x_*)$ (cf. [**GS**]). Also, a singular component v of a harmonic map $u = (V, v) : B_1(0) \to (\mathbf{R}^j, Y_2)$ into a DM-complex satisfies properties (P1), (P2) and (P3) respect to $\mathcal{S} = \mathcal{S}_j(u)$. Similar properties also hold for the singular component of a harmonic map in or into the Weil-Petersson completion Teichmuller space.

LEMMA 72. *Let $v : B_{\sigma_*}(x_*) \to (Y, d)$ be a map satisfying properties (P1), (P2) and (P3) with respect to closed subset $\mathcal{S} \subset B_{\frac{\sigma_*}{2}}(x_*)$. Let $x \in \mathcal{S}$, $\{v_{\sigma_i}\}$ the blow-up maps of v at x and v_0 as in (P2). If $x_i \in \sigma_i \mathcal{S}$ converges to x_0, then*

$$\liminf_{i \to \infty} Ord^{v_{\sigma_i}}(x_i) \leq Ord^{v_0}(x_0).$$

PROOF. Identify $x = 0$ via normal coordinates and let w_{σ_i} be as in (P2). For i sufficiently large,

$$E^{w_{\sigma_i}}_0(1) \leq E^{v_{\sigma_i}}_0(1) = \frac{E^{v_{\sigma_i}}_0(1)}{I^{v_{\sigma_i}}_0(1)} = \frac{\sigma_i E^v_x(\sigma_i)}{I^v_x(\sigma_i)} < 2 Ord^v(x).$$

Thus, for $R \in (0,1)$, [**KS1**] Theorem 2.4.6 implies that $\{w_{\sigma_i}|_{B_R(0)}\}$ has a uniform Lipschitz bound. We can therefore apply lower semicontinuity of energy (cf. [**KS2**] Lemma 3.8) to conclude that, for $x_0 \in B_1(0)$ and $r > 0$ such that $B_r(x_0)$ is compactly contained in $B_1(0)$, we have $E^{v_0}_{x_0}(r) \leq \liminf_{l \to \infty} E^{w_{\sigma_i}}_{x_0}(r)$. On the other hand, by [**KS2**] Theorem 3.9 there is no loss of energy, i.e $E^{v_0}_{x_0}(r) = \lim_{i \to \infty} E^{w_{\sigma_i}}_{x_0}(r)$. By the uniform Lipschitz continuity and the convergence $x_i \to x_0$, we also have $|E^{w_{\sigma_i}}_{x_0}(r) - E^{w_{\sigma_i}}_{x_i}(r)| \leq C|x_i - x_0|$ for some C indepedent of i. Furthermore, (P3) implies that $|E^{v_{\sigma_i}}_{x_i}(r) - E^{w_{\sigma_i}}_{x_i}(r)| < C\sigma_i$. Hence

$$E^{v_0}_{x_0}(r) = \lim_{l \to \infty} E^{v_{\sigma_i}}_{x_i}(r).$$

Furthermore,

$$I^{v_0}_{x_0}(r) = \lim_{i \to \infty} I^{v_{\sigma_i}}_{x_i}(r)$$

by the local uniform convergence in the pullback sense. Combining the above two equalities, we obtain

$$\frac{r E^{v_0}_{x_0}(r)}{I^{v_0}_x(r)} = \lim_{l \to \infty} \frac{r E^{v_{\sigma_i}}_{x_i}(r)}{I^{v_{\sigma_i}}_{x_i}(r)}. \tag{170}$$

Now we apply the monotonicity assumption of (P1) to obtain the result. Indeed, (P1) implies for $c = C + C_1$

$$Ord^{v_{\sigma_i}}(x_i) \leq e^{cr} \frac{r E^{v_{\sigma_i}}_{x_i}(r)}{I^{v_{\sigma_i}}_{x_i}(r)}.$$

Taking liminf as $i \to \infty$ (170) implies
$$\liminf_{i \to \infty} Ord^{v_{\sigma_i}}(x_i) \le e^{cr} \frac{rE_{x_0}^{v_0}(r)}{I_{x_0}^{v_0}(r)}.$$
Finally, by letting $r \to 0$, we obtain the assertion. $\square$

DEFINITION 73. We say that a map $v : B_{\sigma_\star}(x_\star) \to (Y, d)$ satisfying properties (P1), (P2) and (P3) with respect to closed subset $\mathcal{S} \subset B_{\frac{\sigma_\star}{2}}(x_\star)$ satisfies *an order gap property* with respect to $\mathcal{S}$ if there exists $\epsilon_0 > 0$ such that for any $x \in \mathcal{S}$, either $Ord^v(x) = 1$ or $Ord^v(x) \ge 1+\epsilon_0$ (or equivalently, $Ord^{v_0}(0) = 1$ or $Ord^{v_0}(0) \ge 1+\epsilon_0$ for v_0 as in (P2).)

DEFINITION 74. A *higher order point of* v is a point x such that $Ord^v(x)$ exists and is > 1. We denote the set of higher order points of v by $\mathcal{S}_0(v)$.

LEMMA 75. *Let* $v : B_{\sigma_\star}(x_\star) \to (Y,d)$ *be a map satisfying properties (P1), (P2) and (P3) with respect to* $\mathcal{S} \subset B_1(0)$. *If* v *satisfies the order gap property with respect to* $\mathcal{S}$ *as in Definition 73 and* $x \in \mathcal{S}$, $\{v_{\sigma_i}\}$ *and* v_0 *are as in (P2), then*
$$\limsup_{i \to \infty} \dim_\mathcal{H}(\sigma_i^{-1}(\mathcal{S}_0(v) \cap \mathcal{S})) \le \dim_\mathcal{H}(\mathcal{S}_0(v_0)).$$

PROOF. Identify $x = 0$ via normal coordinates. By Lemma 69, it suffices to prove
$$x_i \in \sigma_i^{-1}(\mathcal{S}_0(v) \cap \mathcal{S}) \text{ and } x_i \to x_0 \Rightarrow x_0 \in \mathcal{S}_0(v_0).$$
Since $1 + \epsilon_0 \le Ord^v(\sigma_i x_i) = Ord^{v_{\sigma_i}}(x_i)$ by the order gap assumption, we have $1 + \epsilon_0 \le Ord^{v_0}(x_0)$ by Lemma 72. Hence $x_0 \in \mathcal{S}_0(v_0)$. $\square$

LEMMA 76. *Let* $v : B_{\sigma_\star}(x_\star) \to (Y, d)$ *be a map satisfying properties (P1), (P2) and (P3) with respect to closed subset* $\mathcal{S} \subset B_{\frac{\sigma_\star}{2}}(x_\star)$. *If* v *satisfies the order gap property with respect to* $\mathcal{S}$ *as in Definition 73, then for every* $x \in \mathcal{S}_0(v)$
$$\dim_\mathcal{H}(\mathcal{S}_0(v) \cap \mathcal{S}) \le \dim_\mathcal{H}(\mathcal{S}_0(v_0))$$
where v_0 *is the limit of the blow-up maps of* v *at* x *as in (P2).*

PROOF. Suppose on the contrary that $\dim_\mathcal{H}(\mathcal{S}_0(v) \cap \mathcal{S}) > \dim_\mathcal{H}(\mathcal{S}_0(v_0) \cap \mathcal{S})$ and choose
$$\dim_\mathcal{H}(\mathcal{S}_0(v) \cap \mathcal{S}) > s > \dim_\mathcal{H}(\mathcal{S}_0(v_0)).$$
Since $\mathcal{H}^s(\mathcal{S}_0(v) \cap \mathcal{S}) > 0$, [**Fe**] 2.10.19 implies that there exists $x \in \mathcal{S}_0(v)$ such that (after identifying $x = 0$ via normal coordinates)
$$\lim_{i \to \infty} \mathcal{H}^s(\sigma_i^{-1}(\mathcal{S}_0(v) \cap \mathcal{S})) = \lim_{i \to \infty} \frac{\mathcal{H}^s(\mathcal{S}_0(v) \cap \mathcal{S} \cap B_{\sigma_i}(0))}{\sigma_i^s} \ge 2^{-s}.$$
Thus, $\dim_\mathcal{H}(\sigma_i^{-1}(\mathcal{S}_0(v) \cap \mathcal{S})) \ge s$ for i sufficiently large. By Lemma 75,
$$\dim_\mathcal{H}(\mathcal{S}_0(v_0)) \ge s$$
which is a contradiction. $\square$

DEFINITION 77. Let $v : B_{\sigma_\star}(x_\star) \to (Y, d)$ be a map satisfying properties (P1), (P2) and (P3) with respect to closed subset $\mathcal{S} \subset B_{\frac{\sigma_\star}{2}}(x_\star)$. The map v is said to satisfy the *codimension 2 property of the tangent map* with respect to $\mathcal{S}$ if for any $x \in \mathcal{S}$ and for v_0 the limit of the blow-up maps of v at x as in (P2), we have
$$\dim_\mathcal{H}(\mathcal{S}_0(v_0)) \le n - 2.$$

THEOREM 78. *Let $v : B_{\sigma_\star}(x_\star) \to (Y,d)$ be a map satisfying properties (P1), (P2) and (P3) with respect to $\mathcal{S} \subset B_{\frac{\sigma_\star}{2}}(x_\star)$. If v also satisfies the order gap property with respect to $\mathcal{S}$ as in Definition 73 and the codimension 2 property of the tangent map with respect to $\mathcal{S}$ as in Definition 77, then*

$$\dim_{\mathcal{H}}(\mathcal{S}_0(v) \cap \mathcal{S}) \leq n - 2.$$

PROOF. Since v satisfies the order gap property, we can choose $x \in \mathcal{S}_0(v)$ as in Lemma 76 such that

$$\dim_{\mathcal{H}}(\mathcal{S}_0(v) \cap \mathcal{S}) \leq \dim_{\mathcal{H}}(\mathcal{S}_0(v_0))$$

where v_0 as (P2). The assumption that v satisfies the codimension 2 property of the tangent map implies $\dim_{\mathcal{H}}(\mathcal{S}_0(v_0)) \leq n - 2$. □

Bibliography

[DM1] G. Daskalopoulos and C. Mese, *Harmonic maps between singular spaces I*, Comm. Anal. Geom. **18** (2010), no. 2, 257–337, DOI 10.4310/CAG.2010.v18.n2.a2. MR2672235 (2011g:58027)

[DMV] G. Daskalopoulos, C. Mese, and A. Vdovina, *Superrigidity of hyperbolic buildings*, Geom. Funct. Anal. **21** (2011), no. 4, 905–919, DOI 10.1007/s00039-011-0124-9. MR2827014 (2012g:53071)

[Fe] H. Federer, *Geometric measure theory*, Die Grundlehren der mathematischen Wissenschaften, Band 153, Springer-Verlag New York Inc., New York, 1969. MR0257325 (41 #1976)

[GS] M. Gromov and R. Schoen, *Harmonic maps into singular spaces and p-adic superrigidity for lattices in groups of rank one*, Inst. Hautes Études Sci. Publ. Math. **76** (1992), 165–246. MR1215595 (94e:58032)

[Jo] J. Jost, *Nonpositive curvature: geometric and analytic aspects*, Lectures in Mathematics ETH Zürich, Birkhäuser Verlag, Basel, 1997. MR1451625 (98g:53070)

[KS1] N. J. Korevaar and R. M. Schoen, *Sobolev spaces and harmonic maps for metric space targets*, Comm. Anal. Geom. **1** (1993), no. 3-4, 561–659. MR1266480 (95b:58043)

[KS2] N. J. Korevaar and R. M. Schoen, *Global existence theorems for harmonic maps to non-locally compact spaces*, Comm. Anal. Geom. **5** (1997), no. 2, 333–387. MR1483983 (99b:58061)

[Me] C. Mese, *Harmonic maps into spaces with an upper curvature bound in the sense of Alexandrov*, Math. Z. **242** (2002), no. 4, 633–661, DOI 10.1007/s002090100372. MR1981191 (2004j:58013)

Editorial Information

To be published in the *Memoirs*, a paper must be correct, new, nontrivial, and significant. Further, it must be well written and of interest to a substantial number of mathematicians. Piecemeal results, such as an inconclusive step toward an unproved major theorem or a minor variation on a known result, are in general not acceptable for publication.

Papers appearing in *Memoirs* are generally at least 80 and not more than 200 published pages in length. Papers less than 80 or more than 200 published pages require the approval of the Managing Editor of the Transactions/Memoirs Editorial Board. Published pages are the same size as those generated in the style files provided for $\mathcal{AMS}$-LaTeX or $\mathcal{AMS}$-TeX.

Information on the backlog for this journal can be found on the AMS website starting from http://www.ams.org/memo.

A Consent to Publish is required before we can begin processing your paper. After a paper is accepted for publication, the Providence office will send a Consent to Publish and Copyright Agreement to all authors of the paper. By submitting a paper to the *Memoirs*, authors certify that the results have not been submitted to nor are they under consideration for publication by another journal, conference proceedings, or similar publication.

Information for Authors

Memoirs is an author-prepared publication. Once formatted for print and on-line publication, articles will be published as is with the addition of AMS-prepared frontmatter and backmatter. Articles are not copyedited; however, confirmation copy will be sent to the authors.

Initial submission. The AMS uses Centralized Manuscript Processing for initial submissions. Authors should submit a PDF file using the Initial Manuscript Submission form found at www.ams.org/submission/memo, or send one copy of the manuscript to the following address: Centralized Manuscript Processing, MEMOIRS OF THE AMS, 201 Charles Street, Providence, RI 02904-2294 USA. If a paper copy is being forwarded to the AMS, indicate that it is for *Memoirs* and include the name of the corresponding author, contact information such as email address or mailing address, and the name of an appropriate Editor to review the paper (see the list of Editors below).

The paper must contain a *descriptive title* and an *abstract* that summarizes the article in language suitable for workers in the general field (algebra, analysis, etc.). The *descriptive title* should be short, but informative; useless or vague phrases such as "some remarks about" or "concerning" should be avoided. The *abstract* should be at least one complete sentence, and at most 300 words. Included with the footnotes to the paper should be the 2010 *Mathematics Subject Classification* representing the primary and secondary subjects of the article. The classifications are accessible from www.ams.org/msc/. The Mathematics Subject Classification footnote may be followed by a list of *key words and phrases* describing the subject matter of the article and taken from it. Journal abbreviations used in bibliographies are listed in the latest *Mathematical Reviews* annual index. The series abbreviations are also accessible from www.ams.org/msnhtml/serials.pdf. To help in preparing and verifying references, the AMS offers MR Lookup, a Reference Tool for Linking, at www.ams.org/mrlookup/.

Electronically prepared manuscripts. The AMS encourages electronically prepared manuscripts, with a strong preference for $\mathcal{AMS}$-LaTeX. To this end, the Society has prepared $\mathcal{AMS}$-LaTeX author packages for each AMS publication. Author packages include instructions for preparing electronic manuscripts, samples, and a style file that generates the particular design specifications of that publication series. Though $\mathcal{AMS}$-LaTeX is the highly preferred format of TeX, author packages are also available in $\mathcal{AMS}$-TeX.

Authors may retrieve an author package for *Memoirs of the AMS* from www.ams.org/journals/memo/memoauthorpac.html or via FTP to ftp.ams.org (login as anonymous, enter your complete email address as password, and type cd pub/author-info). The

AMS Author Handbook and the *Instruction Manual* are available in PDF format from the author package link. The author package can also be obtained free of charge by sending email to `tech-support@ams.org` or from the Publication Division, American Mathematical Society, 201 Charles St., Providence, RI 02904-2294, USA. When requesting an author package, please specify $\mathcal{AMS}$-LaTeX or $\mathcal{AMS}$-TeX and the publication in which your paper will appear. Please be sure to include your complete mailing address.

After acceptance. The source files for the final version of the electronic manuscript should be sent to the Providence office immediately after the paper has been accepted for publication. The author should also submit a PDF of the final version of the paper to the editor, who will forward a copy to the Providence office.

Accepted electronically prepared files can be submitted via the web at `www.ams.org/submit-book-journal/`, sent via FTP, or sent on CD to the Electronic Prepress Department, American Mathematical Society, 201 Charles Street, Providence, RI 02904-2294 USA. TeX source files and graphic files can be transferred over the Internet by FTP to the Internet node `ftp.ams.org` (130.44.1.100). When sending a manuscript electronically via CD, please be sure to include a message indicating that the paper is for the *Memoirs*.

Electronic graphics. Comprehensive instructions on preparing graphics are available at `www.ams.org/authors/journals.html`. A few of the major requirements are given here.

Submit files for graphics as EPS (Encapsulated PostScript) files. This includes graphics originated via a graphics application as well as scanned photographs or other computer-generated images. If this is not possible, TIFF files are acceptable as long as they can be opened in Adobe Photoshop or Illustrator.

Authors using graphics packages for the creation of electronic art should also avoid the use of any lines thinner than 0.5 points in width. Many graphics packages allow the user to specify a "hairline" for a very thin line. Hairlines often look acceptable when proofed on a typical laser printer. However, when produced on a high-resolution laser imagesetter, hairlines become nearly invisible and will be lost entirely in the final printing process.

Screens should be set to values between 15% and 85%. Screens which fall outside of this range are too light or too dark to print correctly. Variations of screens within a graphic should be no less than 10%.

Any graphics created in color will be rendered in grayscale for the printed version unless color printing is authorized by the Managing Editor and the Publisher. In general, color graphics will appear in color in the online version.

Inquiries. Any inquiries concerning a paper that has been accepted for publication should be sent to `memo-query@ams.org` or directly to the Electronic Prepress Department, American Mathematical Society, 201 Charles St., Providence, RI 02904-2294 USA.

Editors

This journal is designed particularly for long research papers, normally at least 80 pages in length, and groups of cognate papers in pure and applied mathematics. Papers intended for publication in the *Memoirs* should be addressed to one of the following editors. The AMS uses Centralized Manuscript Processing for initial submissions to AMS journals. Authors should follow instructions listed on the Initial Submission page found at www.ams.org/memo/memosubmit.html.

Algebra, to MICHAEL LARSEN, Department of Mathematics, Rawles Hall, Indiana University, 831 E 3rd Street, Bloomington, IN 47405, USA; e-mail: mjlarsen@indiana.edu

Algebraic geometry, to LUCIA CAPORASO, Department of Mathematics and Physics, Roma Tre University, Largo San Leonardo Murialdo, I-00146 Roma, Italy; e-mail: LCedit@mat.uniroma3.it

Algebraic topology, to SOREN GALATIUS, Department of Mathematics, Stanford University, Stanford, CA 94305 USA; e-mail: transactions@lists.stanford.edu

Arithmetic geometry, to TED CHINBURG, Department of Mathematics, University of Pennsylvania, Philadelphia, PA 19104-6395; e-mail: math-tams@math.upenn.edu

Automorphic forms, representation theory and combinatorics, to DANIEL BUMP, Department of Mathematics, Stanford University, Building 380, Sloan Hall, Stanford, California 94305; e-mail: bump@math.stanford.edu

Combinatorics and discrete geometry, to IGOR PAK, Department of Mathematics, University of California, Los Angeles, California 90095; e-mail: pak@math.ucla.edu

Commutative and homological algebra, to LUCHEZAR L. AVRAMOV, Department of Mathematics, University of Nebraska, Lincoln, NE 68588-0130; e-mail: avramov@math.unl.edu

Differential geometry and global analysis, to CHRIS WOODWARD, Department of Mathematics, Rutgers University, 110 Frelinghuysen Road, Piscataway, NJ 08854; e-mail: ctw@math.rutgers.edu

Dynamical systems and ergodic theory and complex analysis, to YUNPING JIANG, Department of Mathematics, CUNY Queens College and Graduate Center, 65-30 Kissena Blvd., Flushing, NY 11367; e-mail: Yunping.Jiang@qc.cuny.edu

Ergodic theory and combinatorics, to VITALY BERGELSON, Ohio State University, Department of Mathematics, 231 W. 18th Ave, Columbus, OH 43210; e-mail: vitaly@math.ohio-state.edu

Functional analysis and operator algebras, to STEFAAN VAES, KU Leuven, Department of Mathematics, Celestijnenlaan 200B, B-3001 Leuven, Belgium; e-mail: stefaan.vaes@wis.kuleuven.be

Geometric analysis, to TATIANA TORO, Department of Mathematics, University of Washington, Box 354350; e-mail: toro@uw.edu

Geometric topology, to MARK FEIGHN, Math Department, Rutgers University, Newark, NJ 07102; e-mail: feighn@andromeda.rutgers.edu

Harmonic analysis, complex analysis, to MALABIKA PRAMANIK, Department of Mathematics, 1984 Mathematics Road, University of British Columbia, Vancouver, BC, Canada V6T 1Z2; e-mail: malabika@math.ubc.ca

Harmonic analysis, representation theory, and Lie theory, to E. P. VAN DEN BAN, Department of Mathematics, Utrecht University, P.O. Box 80 010, 3508 TA Utrecht, The Netherlands; e-mail: E.P.vandenBan@uu.nl

Logic, to ANTONIO MONTALBAN, Department of Mathematics, The University of California, Berkeley, Evans Hall #3840, Berkeley, California, CA 94720; e-mail: antonio@math.berkeley.edu

Number theory, to SHANKAR SEN, Department of Mathematics, 505 Malott Hall, Cornell University, Ithaca, NY 14853; e-mail: ss70@cornell.edu

Partial differential equations, to MARKUS KEEL, School of Mathematics, University of Minnesota, Minneapolis, MN 55455; e-mail: keel@math.umn.edu

Partial differential equations and functional analysis, to ALEXANDER KISELEV, Department of Mathematics, MS-136, Rice University, 6100 Main Street, Houston, TX 77005; e-mail: kisilev@rice.edu

Probability and statistics, to PATRICK FITZSIMMONS, Department of Mathematics, University of California, San Diego, 9500 Gilman Drive, La Jolla, CA 92093-0112; e-mail: pfitzsim@math.ucsd.edu

Real analysis and partial differential equations, to WILHELM SCHLAG, Department of Mathematics, The University of Chicago, 5734 South University Avenue, Chicago, IL 60615; e-mail: schlag@math.uchicago.edu

All other communications to the editors, should be addressed to the Managing Editor, ALEJANDRO ADEM, Department of Mathematics, The University of British Columbia, Room 121, 1984 Mathematics Road, Vancouver, B.C., Canada V6T 1Z2; e-mail: adem@math.ubc.ca

SELECTED PUBLISHED TITLES IN THIS SERIES

1122 Volodymyr Nekrashevych, Hyperbolic Groupoids and Duality, 2015
1121 Gaëtan Chenevier and David A. Renard, Level One Algebraic Cusp Forms of Classical Groups of Small Rank, 2015
1120 Robert C. Dalang and Marta Sanz-Solé, Hitting Probabilities for Nonlinear Systems of Stochastic Waves, 2015
1119 Joonil Kim, Multiple Hilbert Transforms Associated with Polynomials, 2015
1118 R. Bruggeman, J. Lewis, and D. Zagier, Period Functions for Maass Wave Forms and Cohomology, 2015
1117 Chih-Yun Chuang, Ting-Fang Lee, Fu-Tsun Wei, and Jing Yu, Brandt Matrices and Theta Series over Global Function Fields, 2015
1116 Paul Seidel, Homological Mirror Symmetry for the Quartic Surface, 2015
1115 Pierre Bieliavsky and Victor Gayral, Deformation Quantization for Actions of Kählerian Lie Groups, 2015
1114 Timothy C. Burness, Soumaïa Ghandour, Claude Marion, and Donna M. Testerman, Irreducible Almost Simple Subgroups of Classical Algebraic Groups, 2015
1113 Nicola Gigli, On the Differential Structure of Metric Measure Spaces and Applications, 2015
1112 Martin Hutzenthaler and Arnulf Jentzen, Numerical Approximations of Stochastic Differential Equations with Non-Globally Lipschitz Continuous Coefficients, 2015
1111 Grigor Sargsyan, Hod Mice and the Mouse Set Conjecture, 2015
1110 Masao Tsuzuki, Spectral Means of Central Values of Automorphic L-Functions for GL(2), 2015
1109 Jonah Blasiak, Ketan D. Mulmuley, and Milind Sohoni, Geometric Complexity Theory IV: Nonstandard Quantum Group for the Kronecker Problem, 2015
1108 Chung Pang Mok, Endoscopic Classification of Representations of Quasi-Split Unitary Groups, 2015
1107 Huaxin Lin, Locally AH-Algebras, 2015
1106 A. Rod Gover, Emanuele Latini, and Andrew Waldron, Poincaré-Einstein Holography for Forms via Conformal Geometry in the Bulk, 2015
1105 Tai-Ping Liu and Yanni Zeng, Shock Waves in Conservation Laws with Physical Viscosity, 2014
1104 Gerhard Hiss, William J. Husen, and Kay Magaard, Imprimitive Irreducible Modules for Finite Quasisimple Groups, 2014
1103 J.-M. Delort, Quasi-Linear Perturbations of Hamiltonian Klein-Gordon Equations on Spheres, 2014
1102 Jianyong Qiao, Julia Sets and Complex Singularities of Free Energies, 2014
1101 Jochen Denzler, Herbert Koch, and Robert J. McCann, Higher-Order Time Asymptotics of Fast Diffusion in Euclidean Space: A Dynamical Systems Approach, 2014
1100 Joel Friedman, Sheaves on Graphs, Their Homological Invariants, and a Proof of the Hanna Neumann Conjecture, 2014
1099 Anthony H. Dooley and Guohua Zhang, Local Entropy Theory of a Random Dynamical System, 2014
1098 Peter Keevash and Richard Mycroft, A Geometric Theory for Hypergraph Matching, 2014
1097 Xiaoye Fu and Jean-Pierre Gabardo, Self-Affine Scaling Sets in $\mathbb{R}^2$, 2015
1096 Raphaël Cerf, Critical Population and Error Threshold on the Sharp Peak Landscape for a Moran Model, 2015

For a complete list of titles in this series, visit the
AMS Bookstore at **www.ams.org/bookstore/memoseries/**.